THE AUTHOR

Michael Cooley was born in Tuam in the west of Ireland in 1934. He was educated at local Catholic schools and later studied engineering in Germany. In industry he specialised in engineering design and gained a PhD in computer-aided design.

Mike Cooley was national president of the Designers' Union in 1971 and a TUC delegate for many years. A design engineer for eighteen years, he was a founder member of the Lucas Aerospace Combine Shop Stewards' Committee and one of the authors of its Plan for Socially Useful Production.

He has lectured at universities in Australia, Europe and the United States. He is currently guest professor at the University of Bremen, and visiting professor at the University of Manchester Institute of Science and Technology. He has written for a variety of publications worldwide including the *Guardian*, the *Listener* and the *New Scientist*. He has produced over forty scientific papers and is author or joint author of eleven books in English and German and has contributed to some thirty-five more. His work has been translated into over twenty languages from Finnish to Japanese. He is an international authority on human-centred computer-based systems and in 1981 was joint winner of the $50,000 Alternative Nobel Prize, which he donated to the Lucas Combine Committee.

Mike Cooley is chairman and director of several manufacturing companies in his capacity as director of technology of the Greater London Enterprise Board. He has been married for twenty-six years to a physics teacher, Shirley, and they have two sons.

ARCHITECT OR BEE?

The Human Price of Technology

Mike Cooley

*New Edition,
with a New Introduction by
Anthony Barnett*

A Tigerstripe Book
THE HOGARTH PRESS
LONDON

A TIGERSTRIPE BOOK
This revised edition
published in 1987 by
The Hogarth Press
Chatto & Windus Ltd
30 Bedford Square, London WC1B 3RP

First edition published by Langley Technical Services 1980
Copyright © Mike Cooley 1980, 1987
Introduction copyright © Anthony Barnett 1987

All rights reserved. No part of this publication may be reproduced, stored in a retrieval system, or transmitted in any form or by any means, electronic, mechanical, photocopying, recording or otherwise, without the prior permission of the publisher.

British Library Cataloguing in Publication Data

Cooley, Mike
Architect or bee? : the human price of
technology. – (Hogarth current affairs).
1. Technology – Social aspects
I. Title
306'.46   T14.5

ISBN 0 7012 0769 8

Photoset by Rowland Phototypesetting Ltd,
Bury St Edmunds, Suffolk
Printed in Great Britain by
Cox & Wyman Ltd,
Reading, Berkshire

*To Ernie Scarbrow and the late Danny Conroy,
secretary and chairman of the Lucas Aerospace
Combine Shop Stewards' Committee,
whose imagination and selfless dedication
was an inspiration to me. As mentors and friends
they epitomised for me all that is best
in the trade-union movement.*

A bee puts to shame many an architect in the construction of its cells; but what distinguishes the worst of architects from the best of bees is namely this. The architect will construct in his imagination that which he will ultimately erect in reality. At the end of every labour process, we get that which existed in the consciousness of the labourer at its commencement.

KARL MARX *Capital*

CONTENTS

Introduction	1
1 Identifying the Problem	7
2 The Changing Nature of Work	25
3 The Human–Machine Interaction	38
4 Competence, Skill and 'Training'	54
5 The Potential and the Reality	71
6 Political Implications of New Technology	87
7 Drawing up the Corporate Plan at Lucas Aerospace	114
8 The Lucas Plan – Ten Years On	139
9 Some Social and Technological Projections	157
References	181
Index	186

INTRODUCTION

This book was published out of their back room in 1979 by Mike and Shirley Cooley in response to demands that Mike's argument, in talks, speeches and articles, be made more permanently available. It became a minor legend in its own time. In Britain, over 7000 copies of the first edition of *Architect or Bee?* sold through informal networks. A German translation has now sold 20,000 copies; there have been Swedish, Australian and American editions. In the USA almost every innovative thinker on the social aspects of technology and design cites Cooley. Readers of this new edition will see immediately one of the reasons for the book's appeal. Mike Cooley addresses one of the most central and most difficult issues of our time in a fashion that is completely accessible to people who have no technical knowledge whatsoever.

I first heard of one of the book's chief ideas long before I saw the book itself. This idea whispered, 'It can be done.' It was an idea that said, there is still a *reason* to hope. Although it is not often acknowledged, hope thrives on reason – and on experience. This was another part of the legend (and the truth): the author was an engineer, a scientist with shop-floor credentials; he was saying something practical. The argument that I heard by word of mouth and that warmed the embers of my hope was simplified by its transmission. New technology, said the grapevine, *can* be used to increase the skill of workers without decisive loss of productivity. Therefore the tremendous progress in control and communication, whether globally or in terms of plant and automation, need not turn us into its helpless collaborators. Instead, modern gadgetry can be used to let us be masters of our own destiny.

The argument is immensely attractive because it appeals to a desire that it is almost impossible to express in contemporary capitalism without sounding soft in the head. It refuses to accept as inevitable the division of life into separate spheres of consumption and production, of leisure and work. Cooley insists that modern production itself can at last be made interesting. That in addition to the grind (and we all need a bit of grind) there could be skilled purpose, hence some fulfilment not just in privileged middle-class jobs but also in the factories.

This is quite complicated for a whisper, I realise! I am trying to describe in words a cross between a flash of inspiration and a simple mechanical proposition such as how to use a screwdriver. The way it came across was this: 1) new technology can be very dangerous but it can also be a force for good; 2) its terrific power can be turned to our advantage so that we can *do* more and do it better; 3) this is not just the thought of a kind-hearted left-wing professor but of a practical engineer supported by shop-floor workers at a large plant.

Like all ideas that make you say 'Really?', this one counters a dominant notion: in this case the notion that most of us are condemned to be powerless victims of new technology. Instead, Cooley argues: 1) do not be Luddite about electronic advances; 2) respect and fear its capacity to turn us into vegetables but do not believe that this is inherent in the new technology or even inevitable; 3) seek instead to turn the machines against their present masters, indeed this will be crucial to escape subordination. The attraction of such an argument is obvious. It is modern without being modern*ist*. It is not 'old-fashioned', a fatal characteristic in an age of style, yet it upholds long-standing values which refuse to bend to fashion.

How did my enthusiasm for this idea stand up in the face of the book itself? First I was impressed by the intensity of the critique of automation, new technology and 'Taylorism' – the approach named after the American who first developed time-and-motion studies and applied them to the workplace. By the time I read *Architect or Bee?* I had already heard Mike Cooley talking about 'new technology networks' and devices designed for 'socially useful production', and so I witnessed his passion for machines. The book brings out the fact that he has a much greater passion for human beings, their relationships with each other and the world.

One of the book's qualities is its far-sighted comprehension of

Introduction

the effects of new technology as implemented by present policies. Cooley describes the way computer-controlled machines almost steal the skill of the best workers, to deskill those who will follow into preordained routines. He launches an attack, at once savage and meticulous, on the effects of electronically controlled division of labour. In an interesting new passage in Chapter 4 of this edition he argues that the deskilling of workers is accompanied by the deskilling of consumers. If you fear that you are being turned into a bee in the huge hive of industry with its modern cities and their laid-out suburbs and electronically supervised security, or, even worse, a drone, then here is a book to confirm many of those fears.

Yet this is not a pessimist's essay. Cooley's grasp of the cunning and force that deprives people of their skill and self-confidence stems from his observation of what factories do, while his anger stems from his knowledge that machinery can be organised quite differently. To sustain this critique Cooley mounts a serious attack on the division of labour, obviously, and also on the division of knowledge and authority. He is fond of quoting a Chinese saying:

> I hear and I forget,
> I see and I remember,
> I do and I understand.

His point is not limited to the assertion that we learn by doing. He takes it further; those who know what they are doing – craftspeople with skill and experience – are the ones who best understand what needs to be done.

There is a stunning example at the fulcrum of this book. Cooley was working in Lucas Aerospace in 1975 when 4000 redundancies loomed for its work force of 18,000. The Shop Stewards' Combine Committee drew up a letter which outlined the skills and the capacities of the company's employees and sent it to 180 authorities in the trade unions, in universities and institutions, who had concerned themselves with the problems of structural redeployment. The letter asked them, with these resources what should a company like ours do? What should we be making? Put on the spot, the authorities had no concrete answers. Then the stewards put the same question to their fellow workers, who responded with

150 ideas for products that Lucas Aerospace might make, some of which are described in Chapter 7.

In Sweden Mike Cooley was made joint winner of the Alternative Nobel Prize in 1981, and his award of $50,000 went to the Lucas Committee. Nonetheless Lucas Aerospace would have none of these ideas. Despite the protests from the work force the management forced Cooley out of his job as a senior design engineer. Furthermore, his national union leadership connived in the removal of this 'troublemaker' with his new-fangled proposals and hostility to bureaucracy. The redundancies went ahead, the new products were not made.

It is not my task in an introduction to try and defend or elaborate an author's thesis. But it may help if I describe its originality. Cooley is a socialist. But his argument is an anti-Leninist one in so far as Leninism insisted on the need to introduce correct doctrine into the masses from outside. Lenin, indeed, keenly advocated Taylorism in the USSR as the scientific way to organise work. And yet Cooley is not a 'workerist', who thinks that 'workers know best' as if they already have the correct strategy instinctively inscribed upon their hearts. Someone who praises 'the indispensable advantages of mathematics', who insists on the progressive nature of intensive training schemes and who advocates long apprenticeships can hardly be confused with those who 'believe' in the spontaneous reactions of the underdog. Cooley is a new sort of socialist.

In 1981 the new Labour administration of London under the leadership of Ken Livingstone made Cooley the head of the technology division of the Greater London Enterprise Board. From this vantage point he has been able to set up new technology networks in the British capital and he has helped to initiate a Europe-wide, EEC-funded £3.8 million ESPRIT project to develop a computer-integrated but human-centred manufacturing system. Thus his experience has widened considerably since the first edition, and this new edition of *Architect or Bee?* has been both rewritten in parts and expanded by whole sections to incorporate new material. In the discussion of local government this includes some sharp political reflection on the young leftists with little but their own arrogance to recommend them, who found themselves

Introduction

'in charge' of those whose interests they claimed to represent. There is a blistering account of a 'training adviser' who 'designed a building course to produce a builder in one year' when she knew nothing of the building industry. The account is even more blistering when you hear Cooley talk in private and in detail, free from the risk of libel action, about this and similar incidents.

Cooley, then, is a socialist who offers no easy answers but has a simple and hard-headed sense of direction. This is what impressed me most when I read the book. In particular he believes in work; in the experience that applied effort can bring; in skill, and the thought that it contains; and above all in the human character of working with purpose. For Cooley, 'the future is not "out there" in the sense that a coastline is out there before somebody goes to discover it.' It has yet got to be built by human beings.

ANTHONY BARNETT

One
IDENTIFYING THE PROBLEM

Architect or Bee? was not really written as a book. It is more a mosaic of sketches and views, experience and analysis, worked out in practice and brought together on a rather unusual journey through sections of the engineering industry, trade unions, academic circles and political activities. Obviously, a closely argued conference paper with appropriate references is a very different matter from sections of a speech made from a plinth in Trafalgar Square, yet both are important aspects of the formation of the ideas in this book, and so both are contained here. Inevitably this means that the book is somewhat uneven, but it is based on actual experience – and there is nothing more uneven than the real world!

In spite of this unevenness there are, I believe, consistent threads running through the whole book. Firstly, an assertion that we must always put people before machines, however complex or elegant the machines might be, and, secondly, a sense of marvel and delight at the ability and ingenuity of human beings. I also hope that it will offer some insight into the way we work, and through our work the way we relate to each other and to nature.

It is not enough to identify problems clearly, sharply and sometimes polemically. We also have a profound responsibility to try to do something about them. I seek to be constructive.

Architect or Bee? starts with a critique of the technologies emerging out of the 1960s, and goes on to illustrate the way these concerns found expression in the Lucas Workers' Plan of 1976. This in turn laid the basis for further developments, including technology work at the Greater London Enterprise Board, popular planning in the GLC from 1983 to 1986, and human-centred technologies such as the EEC ESPRIT project, which started in May 1986. I would also like to think that in describing these projects the book highlights some of the problems associated with our top-heavy political structures and their total inability to respond to creative energy from below.

The very first issue to tackle is our overweening faith in science and in technological change. Science is a shallow and arid soil in which to transplant the sensitive and precious roots of our humanity. Faith is indeed the correct word to use in this context. Science and technology are now leading edges in society in rather the same way religion was in medieval times. Furthermore, the zealots of science and technology display much of the missionary zeal of the colonial era. Those who do not understand or, more particularly, accept the dictates of science and technology are almost viewed as lost souls who must be redeemed from their appalling ignorance and, if they cannot be redeemed, are sacrificed at the stake of unemployability.

Nations that show some reluctance to accept the forms of technology developed by the vast multinational corporations are seen as displaying a sort of 'savage ignorance', and denying themselves the earthly heaven which advanced technology could provide for them.

Cultures that are not centred on science and technology are perceived as heresies and clearly need to be exorcised in case they undermine the true faith. Since the new technocratic religion is, by its own definition, 'good', it follows that we should all accept it, and if we are not willing to do so, it is to be imposed upon us 'for our own good'.

Third World countries, which do not want or cannot afford these types of technologies, are held to be 'underdeveloped' not only in the material sense (which they are) but also in a cultural and ideological sense since they lack the understanding and acceptance of the values of the multinational corporations and the technologies that have developed in the nations which are regarded as 'advanced'.

What follows is not a tirade against these forms of technology, but rather a suggestion that we should look sensitively at those forms of science and technology which meet our cultural, historical and societal requirements, and develop more appropriate forms of technology to meet our long-term aspirations.

Identifying the Problem

TECHNOLOGICAL CHANGE

There is still a widespread belief that automation, computerisation and the use of robotic devices will free human beings from soul-destroying, routine, backbreaking tasks and leave them free to engage in more creative work. It is further suggested that this is automatically going to lead to a shorter working week, longer holidays and more leisure time – that it is going to result in 'an improvement in the quality of life'. It is usually added, as a sort of occupational bonus, that the masses of data we will have available to us from computers will make our decisions much more creative, scientific and logical, and that as a result we will have a more rational form of society.[1]

I want to question some of these assumptions, and attempt to show that we are beginning to repeat in the field of intellectual work most of the mistakes already made in the field of skilled manual work at an earlier historical stage when it was subjected to the use of high-capital equipment. I stress the similarity between manual and intellectual work quite deliberately because I resent the division between the two and I will therefore draw parallels between them. Consequently, I look critically at technological change as a whole in order to provide a framework for questioning the way computers are used today.

In my view it would be a mistake to regard the computer as an isolated phenomenon. It is necessary to see it as part of a technological continuum discernible over the last 400 years or so. I see it as another means of production and as such it has to be viewed in the context of the political, ideological and cultural assumptions of the society that has given rise to it.

It seems to me not at all surprising, given the questions we have asked of science and technology, and the 'problems' we have used them to solve, that we now end up with the kind of systems we see all around us. I hold that we have been asking the wrong questions and therefore we have come up with the wrong answers. It is, however, extremely difficult for the public at large to intervene in this process since, instead of baffling them with Latin, the new religion confuses them with mathematics and scientific jargon.

They are led to believe that there is something great and profound going on out there, and it is their own fault that they don't understand it. If only they had a PhD in computer science or theoretical physics they would be able to grapple with the new theological niceties. The scientific language, the symbols, the mathematics and the apparent rationality bludgeon ordinary people's common sense. A concern that things simply are not right and could and should be otherwise is flattened into abject silence.

However, those who do have the appropriate 'qualifications' are also increasingly uncertain, confused and disoriented. The discussions among physicists about the limits of their existing 'objective' techniques and the concern among computer scientists about the implications of artificial intelligence all indicate that the fortress of science and technology in its present form is beginning to show gaping cracks.

Above all this, there is a seething unhappiness among both manual and intellectual workers because the resultant systems tend to absorb the knowledge from them, deny them the right to use their skill and judgement, and render them abject appendages to the machines and systems being developed. Those who are not directly involved in using the equipment are merely confused bystanders. I find a deep concern that individuals feel frustrated because their common sense and knowledge, and their practical experience, whether as a skilled worker, a designer, a mother, a father, a teacher or a nurse, are less and less relevant and are almost an impediment to 'progress'.[2]

Hopefully, we can examine the nature of this 'progress' and seek to identify alternatives which would constitute real progress and involve masses of ordinary people in the definition and construction of that progress.

COMMON SENSE AND TACIT KNOWLEDGE

I will refer frequently throughout the book to 'common sense'. In some respects this is a serious misnomer. Indeed, it may be held to be particularly uncommon. What I mean is a sense of what is to be done and how it is to be done, held in common by those who will

Identifying the Problem

have had some form of apprenticeship and practical experience in the area.

This craftsman's common sense is a vital form of knowledge which is acquired in that complex 'learning by doing' situation which we normally think of as an apprenticeship in the case of manual workers, or perhaps practice in law or medicine.

I shall also refer frequently to tacit knowledge. This knowledge is likewise acquired through doing, or 'attending to things'.

These considerations are of great importance when we consider which forms of computerised systems we should regard as acceptable.

It is said that we are now approaching, or are actually in, an information society. This is held to be so because we are said to have around us 'information systems'. Most of such systems I encounter could be better described as data systems. It is true that data suitably organised and acted upon may become information. Information absorbed, understood and applied by people may become knowledge. Knowledge frequently applied in a domain may become wisdom, and wisdom the basis for positive action.

All this may be conceptualised as at Figure 1 in the form of a noise-to-signal ratio. There is much noise in society, but the signal is frequently dimmed.

Another way of viewing it would be the objective as compared with the subjective.

At the data end, we may be said to have calculation; at the wisdom end, we may be said to have judgement. Throughout, I shall be questioning the desirability of basing our design philosophy on the data/information part rather than on the knowledge/wisdom part. It is at the knowledge/wisdom part of the cybernetic loop that we encounter this tacit knowledge to which I will frequently refer.

The interaction between the subjective and the objective, as indicated in Figure 2, is of particular importance when we consider the design of expert systems. In this context, I hold a skilled craftsworker to be an expert just as much as I hold a medical practitioner or a lawyer to be an expert in those areas.

If we regard the total area of knowledge required to be an expert

Architect or Bee?

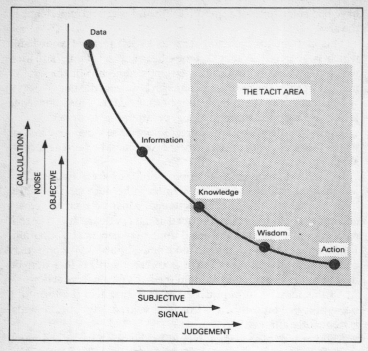

Fig. 1. The tacit area.

as that represented by A, we will find that within it there is a core of knowledge (B) which we may refer to as the facts of the domain, the form of detailed information to be found in a text book.

The area covered by B can readily be reduced to a rule-based system. The annulus AB may be said to represent heuristics, fuzzy reasoning, tacit knowledge and imagination. I hold that well-designed systems admit to the significance of that tacit knowledge and facilitate and enhance it. I reject the notion that the ultimate objective of an expert system should be so to expand B that it totally subsumes A. It is precisely that interaction between the objective and the subjective that is so important, and it is the concentration upon the so-called objective at the expense of the subjective that is

Identifying the Problem

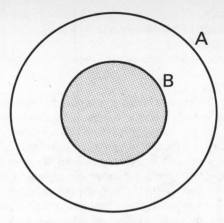

Fig. 2. The limits of rule-based systems.

the basis of the concern expressed in respect of existing systems design.

THE ACQUISITION OF SKILL

In the processes and systems described below, my concern is not merely about production but also about the reproduction of knowledge. I frequently refer to learning by doing, for as a result of this, human beings acquire 'intuition' and 'know-how' in the sense in which Dreyfus uses these. This is not in contradiction with Polanyi's concept of tacit knowledge; it is rather a description of a dynamic situation in which through skill acquisition people are capable of integrating analysis and intuition. Dreyfus and Dreyfus[3] distinguished five stages of skill acquisition: 1) novice; 2) advanced beginner; 3) competent; 4) proficient; and 5) expert.

I think learning-development situations are absolutely vital, and when someone has reached the knowledge/wisdom end of the cybernetic transformation (see Figure 1) and has become an 'expert' in the Dreyfus sense, they are able to recognise whole scenes without decomposing them into their narrow features. Thus I do not counterpose tacit knowledge, intuition or know-how against analytical thinking, but rather believe that a holistic work situation is one which provides the correct balance between analytical thinking and intuition.

Broadly stated, Dreyfus views skill acquisition as follows:

Stage 1 Novice

At this stage, the relevant components of the situation are defined for the novice in such a way as to enable him or her to recognise them without reference to the overall situation in which they occur. That is, the novice is following 'context-free rules'.

The novice lacks any coherent sense of the overall task and judges his or her performance mainly by how well the learned rules are followed. Following these rules, the novice's manner of problem solving is purely analytical and any understanding of the activities and the outcome in relation to the overall task is detached.

Stage 2 Advanced Beginner

Through practical experience in concrete situations the individual gradually learns to recognise 'situational' elements, that is, elements which cannot be defined in terms of objectively recognisable context-free features. The advanced beginner does it by perceiving a similarity to prior examples. The growing ability to incorporate situational components distinguishes the advanced beginner from the novice.

Stage 3 Competence

Through more experience the advanced beginner may reach the competent level. To perform at the competent level requires choosing an organisational plan or perspective. The method of understanding and decision-making is still analytical and detached, though in a more complex manner than that of the novice and the advanced beginner.

The competent performer chooses a plan which affects behaviour much more than the advanced beginner's recognition of particular situational elements, and is more likely to feel responsible for, and be involved in, the possible outcome. The novice and the advanced beginner may consider an unfortunate outcome to be

Identifying the Problem

a result of inadequately specified rules or elements, while the competent performer may see it as a result of a wrong choice of perspectives.

Stage 4 Proficiency

Through more experience, the competent performer may reach the stage of proficiency. At this stage the performer has acquired an intuitive ability to use patterns without decomposing them into component features. Dreyfus calls it 'holistic similarity recognition', 'intuition' or 'know-how'. He uses them synonymously and defines them as 'the understanding that effortlessly occurs upon seeing similarities with previous experiences . . . intuition is the product of deep situational involvement and recognition of similarity'.

Though intuitively organising and understanding a task, the proficient performer is still thinking analytically about how to perform it. The difference between the competent and the proficient performer is that the proficient performer has developed an intuitive way of understanding based on more experience while the competent performer is still forced to rely on the detached and analytical way of understanding the problem.

Stage 5 Expertise

With enough experience, the proficient performer may reach the expert level. At this level, not only situations but also associated decisions are intuitively understood. Using still more intuitive skills, the expert may also cope with uncertainties and unforeseen or critical situations.

Dreyfus and Dreyfus's essential point is to assert that analytical thinking and intuition are not two mutually conflicting ways of understanding or of making judgements. Rather they are seen to be complementary factors which work together but with growing importance centred on intuition when the skilled performer becomes more experienced. Highly experienced people seem to be able to recognise whole scenarios without decomposing them into elements or separate features.

My criticism of the prevailing systems-design methodology and philosophy and my rather scathing remarks about 'training' in Chapter 4 stem from the fact that in both cases they deny us that 'deep situational involvement'. Our development tends to be constrained within the novice end of the skill-acquisition spectrum.

I describe later those experiences, systems and machines which could reverse this approach and provide instead developmental situations to facilitate the acquisition of those attributes to be found at the expert end of the skill spectrum.

Many designers fear to discuss these concerns because they may be accused of being 'unscientific'. There is no suggestion in this line of argument that one should abandon the 'scientific method'; rather we should understand that this method is merely complementary to experience and should not override it, and that experience includes 'experience of self as a specifically and differentially existing part of the universe of reality'.[4] Such a view would help us to escape from the dangers of scientism which, as was once suggested, may be nothing more than a Euro-American disease.[5]

FRAGMENTATION

I take the Hegelian view that truth lies in the totality, and therefore, after considering some of the equipment currently in use, I will relate its effects to the labour process and try to give an overall view of what is happening. The equipment and processes described are not necessarily the most advanced or the latest in their field. They are chosen because they are typical of the kind of changes that are taking place in design.[6] The problems I describe within the design activity can be regarded as universal problems and also apply to computers in insurance, banks, newsprint industry or any other field.

The Equipment

The first system considered is an early forerunner of that which culminated in the mid 1980s in advanced CAD (computer-aided design) and complete CIM (computer-integrated manufacturing) systems. It conveniently serves to demonstrate the tendency to fragment, and ultimately replace, the functions of the draughtsman with computer-based equipment. In Britain and much of Europe up to the 1940s the draughtsman was the centre of the design activity. He could design a component, draw it, stress it out, specify the material for it and the lubrication required. Nowadays, each of these is fragmented down to isolated functions. The designer designs, the draughtsman draws, the metallurgist specifies the material, the stress analyst analyses the structure and the tribologist specifies the lubrication. Each of these fragmented parts can be taken over by equipment such as this automatic draughting equipment. (See Figure 3.)

With this equipment, the draughtsman no longer needs to produce a drawing and so the subtle interplay of interpretation and modification as the commodity was being designed and related to the skilled manual workers on the shop floor is being ruptured. What the draughtsman now does is work on the digitiser and input the material through a graticule or teletype. An exact reading is set of the length of each line, the tolerance and other details.

The design emerges as a set of 'instructions' which are expanded in the computer and then used to control a machine tool such as a jig borer, lathe or continuous path milling machine. The same 'instructions' may also be used to control a device which undertakes the inspection function. If perchance you want a drawing in order to show the customer exactly what they are purchasing – and that's the only reason you would bother to do it – then you can produce one on a master plotter very accurately. You can get a less accurate one on the microplotter, which also produces an aperture card.

What is important in all this is not only that the fragmented functions of the designer have been built into the computer, but that the highly skilled and satisfying work on the shop floor has been destroyed. It is no longer a question of supply and demand, of a slump or a boom; these jobs have been technologically eliminated.

The second system considered is a design system called manned computer graphics. In the past, skilled workers have had a tacit understanding of mathematics through their ability to analyse the size and shape of components by working on them.[7] More and

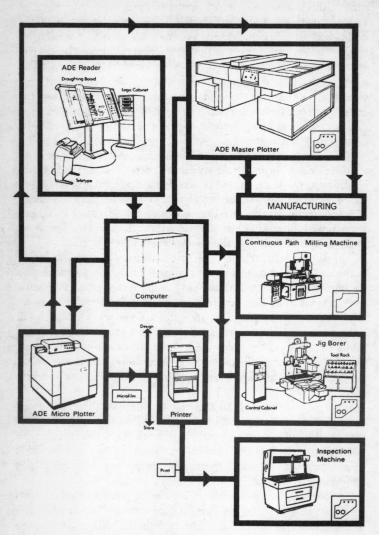

Fig. 3. ADE system for engineering data processing.

Identifying the Problem

more, that knowledge has been abstracted away from the labour process and rarefied into mathematical functions. In diagrammatical form the function might, for example, be of the kind shown in Figure 4. This is a sinusoidal function and might represent the way a shaft is vibrating.

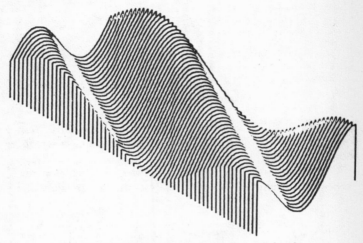

Fig. 4. Computer-produced solution space surface for
SIN $(8*(X-1)/X_L + 1/4\,(Y-1) + 1.0$.

A major application of manned computer graphics is in the field of structural analysis. Equations required for the analysis of the structure are automatically set up and are solved automatically upon request of the analytical output. Displacement, loads, shear and moments are computed and conveniently displayed for perusal. Changes of input conditions are easily facilitated and the corresponding output is displayed upon request. Constraining forces may be placed by using a light pen. Figure 5 shows the exaggerated displacement under load.

This equipment represents a deskilling process because it becomes possible to use designers and stress analysts with much less ability and experience than was previously required.

The windloading on a tower is a quite complex analysis problem.

Architect or Bee?

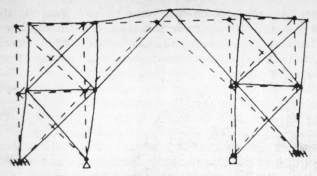

Fig. 5. Exaggerated displacement under load.

The stress in the structure as it distorts can be obtained from a computer package. The distortion is represented on a screen as shown in Figure 6.

What all this means is that the knowledge which previously existed in the mind of the stress analyst, which was taken home

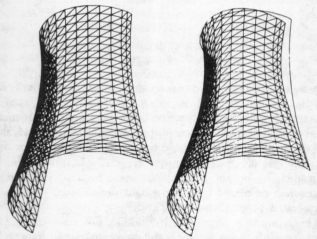

Fig. 6. The illustrations are finite element idealisations of a sectioned cooling tower. The right-hand picture shows the exaggerated distortion under wind loading.

every night and was part of that person's bargaining power, has now been extracted from them. It has been absorbed and objectivised into the machine through the intervention of the computer and is now the property of the employer, so the employer now appropriates a part of the worker and not just the surplus value of the product. Thus we can say that the worker has conferred 'life' on the machine and the more he gives to the machine, the less there is left of himself.

Further Examples

Another possibility for the computer, as some architectural readers will know, is to analyse a whole range of variables in an application and plot the solution. This has been used by a planning agency who required a layout of villas on an island. The layout was to be such that each villa had the same amount of sunlight, garden space, view of the sea and many other variables. The computer handled this by doing an initial layout and gradually rearranging and modifying it to fit in with the terrain until it ended up with a final layout superimposed on the map of the island. It created a very dense distribution of buildings (in my view a grotesque thing to do to an island). This too was very early work. There are much more sophisticated packages available today.

In the medical field there are several uses of computer-aided design which in my view are positive although they bring with them a whole range of problems which I shall describe later.

One example is the use of a visual display unit in the design of equipment for ear protection. The VDU will display the form of soundwaves in the inner ear so your protective equipment can be modified until certain sounds are shown to have been eliminated. Theoretically, you could design ear-protection equipment that would allow human speech through and eliminate other forms of noise. You could in fact choose what you want to hear.

A second example is the use of a VDU in the design of artificial limbs. A graphic system will work out the area of the kneecap joint required for the particular individual for whom it is being designed. The whole structure can be animated on the screen. The person who was to use the limb could be involved in discussing its design before it was actually made. The limb could therefore be designed specifically for the patient unlike those ill-fitting objects which compel the patient to hobble along, often at an angle of ten degrees to the vertical, and still have a stench of Victorian charity about them!

A third example is the use of computers in the design of heart valves. Techniques originally developed to display characteristics in hydraulic circuits in aircraft are used to display the venturi and other characteristics and the blood flowing through the heart valve. Working interactively, it is possible to modify the valve-orifice diameters and other critical physical dimensions and display on the screen the resultant flow characteristics. It is thus possible to optimise the heart-valve design to meet the special requirements of the individual patient.

When one considers all these uses for the computerised equipment, one gets the immediate impression that it must automatically improve the whole creativity of the designer using it. However, there are enormous problems involved which require discussion. The complex communications that go on between human beings during problem-solving activities are being distanced by the computer and by the systems interfacing the people with the computer, and the consequences are very serious and far-reaching. Look at the job of a building designer, for example. In the past, when designing a building, he would go out to the site to see how the structure was progressing. He would discuss it with the site engineer and maybe modify the design. Now it is possible to have a VDU on the site so that visits are unnecessary because the designer and engineer can have a conversation via the equipment. The designer's drawings will be transmitted through British Telecom lines and displayed on the screen. The physical contact between the designer and the site is cut out. Apart from the design implications, the system will tie people down to the machine more and more and the break of getting away from the drawing office and on to the site, which was always one of the perks of the job, will no longer be necessary.

Identifying the Problem

MODELS OF REALITY

Part of the skill of a draughtsman or a designer was the ability to look at a drawing and conceptualise what the product would look like in practice. That conceptualisation process is now also being eliminated by computers. There are systems capable of tracing round the profile of a conventional drawing which includes plan and elevation views, and producing an accurate three-dimensional representation of the object on the screen. The computer will rotate it through any angle for you when given instructions. This can be extended further in the field of architecture, for example. A visual display like the one described could be made of any proposed municipal building, and local people could look at it and see whether they approved of its design and its location. Normally, a plan of a proposed municipal building is available for inspection in the town hall, but to most people this means very little. It is intelligible only to an elite group.

By sensitive use of the computer in this way we could involve the community in deciding the kind of buildings it wanted.[8]

Theoretically, then, there is the potential for democratising the decision-making process. I will argue elsewhere, however, that the computer is in fact used to reinforce the power of minorities over majorities. There is a real danger in that the whole design process could be extremely manipulative. If you have a perspective view of a building on a visual display unit and you take the point of convergence far away, you can make the building look slim and attractive, disappearing into the horizon. On the other hand, if you take it close up you can make the building look like a high-rise block. Thus it is very easy to manipulate public opinion and I think that some architects are not beyond that sort of thing.

At an even higher level you can get what appears to be all the power of retrospective logic. Anyone who has worked as a designer will know that you get your best ideas afterwards, when you can see the mistakes you have made while designing. There are systems now used in the field of architecture which aim to provide the designer with some kind of retrospective logic. They were adapted from visual simulation techniques used to train astronauts in docking manoeuvres. The underlying principle is that images are presented thirty times per second on a colour visual display unit. Standard cues of depth are given as overlapping surfaces and the apparent size of the object is given as inversely proportional to the distance from the observer. This of course is the typical Western cultural way of presenting visual data of this kind.

In computer-aided architectural design, each building and object is defined in its own three-dimensional coordinate system. These are then presented as a hierarchical structure of coordinate data. This means that all the existing buildings can be input as data structure and the new building to be designed is shown within the context of the existing architectural arrangements. The VDU can give the user the illusion of walking towards a building that does not yet exist. One can experience the sensation of going inside the proposed building and looking out at the existing buildings. One can take windows out, move them about, enlarge the entire building and take it right outside the proposed site. The aim is to assess the total effect of the new building on the whole environment before constructing it. There are already grounds for believing, however, that images of reality as presented in that form are still very different from the actuality. When the building is erected you can get a sort of ghettolike prison atmosphere which is not apparent on the screen.

Quite apart from the destruction of the creativity the worker used in doing the job, what must be of concern to all of us is where the next generation of skills is coming from, skills which will need to be embodied in further levels of machines. The feel for the physical world about us is being lost due to the intervention of computerised equipment and work is becoming an abstraction from the real world. In my view, profound problems face us in the coming years due to this process.[5] If human beings increasingly work with models of reality rather than with reality itself, and are thereby denied the precious learning process which flows from it and the accumulation of tacit knowledge, the problems are likely to be significant and have been discussed by writers of widely varying 'political stances'.

Two
THE CHANGING NATURE OF WORK

RATE OF CHANGE

In spite of the power of computer equipment to do some really good work, it brings in its wake all the problems which high-capital equipment brought to manual work at an earlier historical stage. Firstly, it shares with all other equipment historically an ever increasing rate of obsolescence. Wheeled transport existed in its primitive form for thousands of years; Watt's steam engine was working for over 100 years after it was built. High-capital equipment in the 1930s was written off after twenty-five years and equipment of the latest kind will be obsolete in three or four years' time. Economists would say that this shows the increasingly short life of fixed capital.

Further, when viewed historically, it will be seen that the total cost of the means of production is ever increasing. This is in spite of the reduction in the cost of hardware. While these costs are reduced dramatically as computerised systems are miniaturised, the total cost of the system, including the plant and the processes which the hardware is used to control, is increasing. The most complicated lathe one could get 100 years ago would have cost the equivalent of ten workers' wages per annum. Today, a lathe of comparable complexity, with its computer control and the total environment necessary for the preparation of the software and the operation of the machine, will cost something in the order of 50 workers' wages per annum. This is frequently forgotten when people talk about microprocessors, and you are almost given the impression that you could fly the Atlantic on a chip, that you could excavate the foundations of buildings or process chemicals (even food) with microprocessors in isolation.

A discernible feature of modern equipment of any kind is the rate of change that is now driving us along at an incredible tempo.

Architect or Bee?

Over the last century alone, the speed of communication has increased by 10^7, of travel by 10^2, of data handling by 10^6. Over the same period, energy resources have increased by 10^3 and weapon power by 10^6. We are being drawn into a tremendous technological inferno, and it means that the knowledge we have and the basis upon which we judge the world about us are becoming obsolete at an ever increasing rate, just like the equipment. It is now the case in many fields of endeavour that simply to stand still, you have to spend 15 per cent of your time updating your knowledge. The problems for older workers are enormous.

There is a mathematical model justifying this.

Suppose S represents the total stock of useful theoretical knowledge possessed by an engineer.
         F the fraction of this knowledge which becomes obsolete each year.
         R the fraction of his working time devoted to acquiring fresh theoretical knowledge.
         L his learning rate.
Then   LR = FS.

Assuming S is constant and equal to the stock of knowledge with which the engineer left university and that his average rate of learning remains the same as it was during his three-year university course, and assuming also that 5 per cent of his knowledge becomes obsolete each year, the equation becomes:

$$R \frac{S}{3} = 0.05\, S \quad \text{whence } R = 0.15 \text{ or } 15\% \text{ of working time.}$$

As might well be expected, the number of journals to be studied is also increasing. This was shown by Hilary and Steven Rose in 'Science & Society' (Figure 7).

In some fields, the rate of obsolescence is much greater than that indicated above, particularly in certain areas of computer application. As far back as 1972, Norman Macrae, deputy editor of the *Economist*, stated,[1] 'The speed of technological advance has been so tremendous during the past decade that the useful life of the knowledge of many of those trained to use computers has been

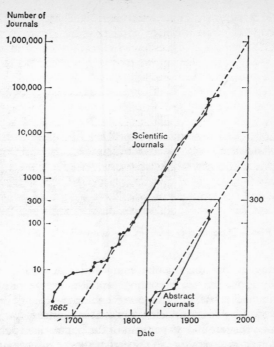

Fig. 7. Total number of scientific journals and abstract journals founded, as a function of date.

about three years.' He further estimated that 'a man who is successful enough to reach a fairly busy job at the age of thirty, so busy that he cannot take sabbatical periods for study, is likely by the age of sixty to have only about one eighth of the scientific (including business scientific) knowledge that he ought to have for proper functioning in his job'.

It has been said that if you could divide knowledge into quartiles of outdatedness, all those over the age of forty would be in the same quartile as Pythagoras and Archimedes. This alone shows the amazing rate of change, and the stress it places on design staff, particularly the older ones, should not be underestimated. What is

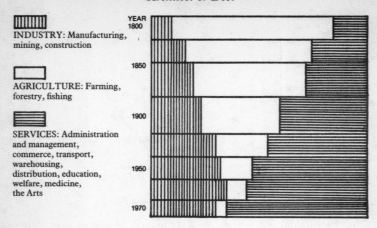

Fig. 8. The changing structure of employment in the USA.

happening is that the organic composition of capital is being changed. Processes are becoming capital-intensive rather than labour-intensive. As a result of technological change we have moved towards a form of society depicted in Figure 8.

It shows that around 86 per cent of the population of the USA was involved in agriculture in the early 1800s. This was subjected to mechanisation, the use of chemicals and then automation, so that now only 6 per cent of the population produces a far greater agricultural output than in 1800. There are automatic tractors that can feel their way round a field so that no human being is required. (It should be pointed out that the calorific value of the food so produced is actually less than the equivalent energy input if one takes into account the tractors, harvesters and chemicals. This is a problem society may have to address in the long term.)

During the same period, manufacturing industry was growing. Up to the 1950s and certainly the early 1960s it was subjected increasingly to mechanisation and automation. The proportion employed in manufacturing is now reduced to about 30 per cent and declining rapidly, while at the same time, the administrative, information and scientific area has been growing. (Figure 9.)

The Changing Nature of Work

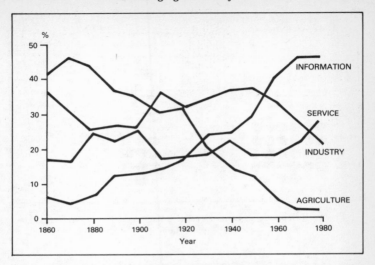

Fig. 9. Change in composition of US work force by percentage, 1860–1980.

That in turn is now being subjected to massive computerisation and automation which will do to this area what has been done to the others. We are confronted therefore with high and growing structural unemployment. More and more we are moving into a position where large numbers of people are being denied the right to work at all.

If we look at given sectors of industries, for example the manufacture of telecommunication exchanges, we will see that whereas it required twenty-six workers to produce one unit of switching power in the electro-mechanical field, with a first-generation electronic system it is ten, and with fully electronic systems in the year 1990 it will require only one worker. (Figure 10.)

Such productivity increases of 26 to 1 cannot be met by increases in production. Our inability to do so results from the constraints of energy and materials.

The fact that there is no work for large numbers of people may

Architect or Bee?

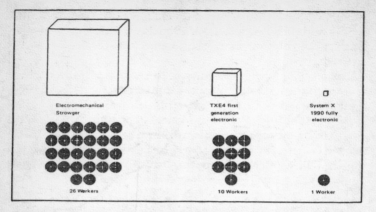

Fig. 10. Telephone exchanges – relative sizes and labour ratios needed to make them.

not seem too great a tragedy to some. Indeed, it has been said that if work were such a good and fulfilling activity, the ruling class would have monopolised it themselves. There is also the view that if people lose their jobs in traditional industries, it will simply liberate them to engage in more creative activities. During one of the closures in the steel industry, a middle-class member of one of the audiences I addressed asserted this so strongly that he must have believed it possible for a redundant steelworker to get a block of wood, carve a Stradivarius, find three of his mates who had done something similar and together play the late string quartets of Beethoven! What we witness is not freedom to enjoy leisure, but rather enforced idleness. The educational, cultural and other facilities simply do not exist in Britain to allow people to enjoy leisure fully, nor are the economic resources available, and leisure can often be a comparatively expensive activity. Furthermore, the cultural background gives no basis for this.

However, over and above this, I do hold that work is vitally important for human beings. I do not mean here grotesque alienated work like that on a production line at Ford's – the type of work developed for the last fifty years – but work in its historical sense which links hand and brain in a meaningful, creative process. I also hold it to be important that work should provide a balance of a range of activities – manual and intellectual, creative and non-creative.

It is precisely these imbalances in forms of modern work that cause people to frantically pursue leisure activities and therapeutic activities on the one hand and artificially contrived forms of exercise on the other. There is something particularly ludicrous about a situation where people will drive into London in their individual cars, sit at a desk all day (denied even the exercise of walking round to collect documents, which they may now call up on their VDUs) and then go home in the evening and got on an exercise bike (where they are not going anyplace) or, more absurdly, on to an exercise spade. These spades have a small microprocessor through which the users can input their weight, age, sex and medical details such as a previous heart attack. An appropriate programme of exercise will be defined. There in the comfort of their own front room they can pretend they are digging. In those circumstances it is leisure and fun. If they were required to do a little digging in the course of their work during the day they would be downright insulted by having to do manual work.

Work also provides a learning, developing situation, and one through which we begin to identify ourselves. If, for example, you ask somebody what they are, they never say, 'I'm a Beethoven lover', 'a James Joyce reader' or even 'a keen footballer'. They will say, 'I'm a fitter', 'a nurse', 'a teacher' and so on.

What is being suggested here is that we need to develop new or holistic forms of work which link the human being as producer and consumer, and which provide a balanced range of intellectual and manual activities, given that these will vary considerably from individual to individual. Furthermore, the hours of work will have to become much more flexible and will also need to be linked with a shorter working week, longer holidays and more leisure time. These, it seems to me, are prerequisites for dealing with the growing structural unemployment which is now beginning to be evident in most of the technologically advanced nations.

It used to be suggested that, as jobs were lost in the manufacturing industries as a result of technological change, new jobs would become available in the white-collar and service sectors, but computerisation is taking its toll there as well. A recent report in France (the NORA report) suggested that modern computing technology in banks over the next ten years will reduce the staff in that area by 30 per cent. Similar figures apply to the insurance industry. A survey in West Germany suggests that by 1990, 40 per cent of the present office work will be carried out by computerised systems. The West German unions have translated this into figures and calculated that it would mean a staggering loss of 2 million out of West Germany's 5 million clerical and office jobs. Thus, not only will there be labour displacement in the manufacturing industries, but those displaced and their children will be in competition with growing numbers of other white-collar workers who have themselves been displaced by these systems.

It is being suggested that even if the EEC countries could maintain their present growth rates there would still be 20 million people out of work by 1988. Two issues arise here. First, we are underestimating how important meaningful work is to human beings.

Secondly, we should recognise that these jobs are not ours to give away. We are the guardians of those jobs for future generations, and in guarding them we can also ensure that there are jobs which are more fulfilling and creative than the ones we have now got. Once society begins to accept that the unemployment problems it is facing are structural rather than cyclical, there should then be the basis for mobilising public opinion, political and other movements to begin to redress the situation.

There is more to it than that, however, for those who are displaced by technological change are not the only ones seriously affected. What is happening to those remaining in work is really worth analysing.

THE PRESSURE IS ON!

It is widely recognised on the shop floor that technological change has resulted in a frantic work tempo for those who remain. Even at the Triumph plant in Coventry in the mid '70s, it was reckoned that a worker was 'burned up' in ten years when working on the main track. The engineering union to which I belonged at the time was asked to agree that nobody would be recruited over the age of thirty, so that the last ten years would be from thirty to forty. The same kind of thing is happening in parts of the steel industry. The workers there have become party to an agreement which includes a medical check.

Now, in a civilised society, a medical check would be an excellent thing. If something were the matter with you it would be discovered, put right, and you would continue working. *This* medical check is a sort of industrial MOT test. Your response rate is worked out (like a diode) to see whether or not you are fast enough at interfacing with the equipment. If you fail, you are put on to second- or third-rate work. There is an established list of wages for those who have been so replaced because their reaction time was not fast enough.

For those who do not work in the automotive industry, it is difficult to appreciate how bad the situation has become and to what extent workers are even being paced by these computerised high-technology systems. In the section where they press out the car bodies in one car company, workers in 1973 were subject to an agreement on the make-up of their rest allowance. The elements are as follows:

Trips to the lavatory	1.62 minutes. It is computer precise; not 1.6 or 1.7 but 1.62!
For fatigue	1.3 minutes
Sitting down after standing too long	65 seconds
For monotony	32 seconds

And so the grotesque litany goes on.[2] The methods engineers located the toilets strategically close to the production line so that operators could literally flash in and flash out. What arrogance some technologist had, to be able to do that to another human being! If we have strikes in the automotive industry we must not be surprised. In my view they are right to strike against conditions of this kind, yet this is the kind of philosophy behind the design of much of the equipment produced for industry today.

TASK-ORIENTED TIME

Compare this itemised industrial agreement with the playwright J. M. Synge's account of the Aran islands, which provides us with a vivid example of differing notions of time and its relation to natural rhythms of work. In such natural rhythms fishing boats are launched to attend the tides, crops must be sown in spring and harvested in autumn, cows have to be milked when their udders are full and sheep guarded during lambing time. 'Few people, however, are sufficiently used to modern time to understand in more than a vague way the convention of the hours. And when I tell them what o'clock it is by my watch, they are not satisfied, and ask how long is left them before the twilight.'

Time viewed in this way is task-oriented, more understandable, acceptable and natural than timed labour. The worker deals with tasks which are comprehensible and universally agreed necessities. In activities organised in this manner, there appears to be less demarcation between work and life, between the human being as producer and as consumer. Social activities like harvest festivals are integral to the process. Social interaction and labour are intermingled during the working day, which extends or contracts according to the season and the work. Time lost because of bad weather on one day will mean labouring until dusk on another. Time is used according to the task in hand and the changing circumstances in which the task will be performed. It is not predetermined synthetically.

Such notions of time apply primarily in rural societies but are relevant also to craftspeople, writers, artists and those not totally subordinated to the industrial machine, where, in some cases, as in the example given above, the tasks of the worker are timed with a ruthless precision and are the source of continual industrial unrest. The counterproductive nature of treating human beings in this way still comes as a surprise to some outside observers. A report from Rome in early 1985 indicated that in a major car-manufacturing company with over 180,000 employees, 147,000 of whom were factory workers, 21,000 were missing on a Monday and there was a daily absentee rate averaging 14,000. Throughout the whole of the Italian economy, an average of 800,000 workers per day are absent out of a total of 20 million, according to a management association report. This was attributed to 'the increasing disgust of younger workers with the assembly-line discipline and the recent influx of untrained [sic] southern Italians into northern factories'.

EMPLOYERS CONSIDER ALTERNATIVES

Less spectacular, but even more significant as indicators, are the rising rate of production defects and errors, the widespread increase in accidents, absenteeism and turnover, and the very real difficulty, in spite of the bait of a financial anaesthetic, of finding adequate numbers of workers to submit to the degradation of the modern factory. New forms of work organisation in Sweden and Japan are indications that employers feel compelled to explore alternatives which provide some dignity and autonomy, however small, for the workers involved.

Even when the employer does succeed in finding sufficient 'human appendages', his problems are by no means at an end. The industrial worker, despite a class-ridden educational system which systematically seeks to reduce his or her expectations to an absolute minimum, and despite the continual bludgeoning by the mass media, still retains a degree of dignity and ingenuity which employers find alarming. Indeed, it is one of the greatest tributes to human dignity that the industrial worker obstinately refuses to meet the specification given by Frederick W. Taylor, 'that he should be so stupid and so phlegmatic that he more nearly resembles in his mental make-up the ox, than any other type'. (Taylor, the originator of 'scientific management', first announced his theories to American engineers in 1895. His system analyses and subdivides work into its smallest mechanical components and work activities and rearranges these elements into the most effective combination. Each component operation is timed by a stopwatch and a standard time is specified for each task. Taylor extended the division of labour into a division of time itself. The stopwatch had been used before Taylor, but only for entire jobs. He used it for each minute activity of the worker in isolation. In Taylorism, the stopwatch and its modern computer-based successors are the Bible.)

It is not surprising that human beings, when viewed in this way and when required to work within a productive process which treats them as oxen, should take what steps they can, however defensive, to assert their humanity. These attempts are not unrelated to the failure rate in parts of industry. It had reached such proportions by 1980 that something like half the equipment lay idle at General Motors' most modern factory where, as predicted by Gorz, the French political theorist, in 1976, 'the intensity and monotony of work surpasses anything previously imposed on assembly-line workers'.

So disgraceful was this waste of human and material resources that employers have felt compelled to accept improved health and safety standards and provide a less alienating environment. However, accidents and illnesses of a physical nature are replaced by psychological and stress factors to such a degree that employers seek aid from industrial psychologists, group technologists and job-enrichment specialists. These advisers in no way change the basic power relationships which give rise to the problems in the first place. It is, as a Lucas shop steward put it, 'like keeping people in a cage and debating with them the colour of the bars'.

A more acceptable solution from the employer's viewpoint, particularly in the case of errors and underutilisation of plant in the automotive industry, is to dramatically reduce the number of people required by replacing them with robots as we move towards the workerless factory. Robotic equipment is becoming cheaper and cheaper each year in comparison to the annual income of a worker employed in the same area. (See Figure 11.)

We may summarise by saying that human beings do need work, even if we redefine and restructure it significantly. Work provides people with an opportunity to express themselves at a number of different levels – to handle uncertainty, to cope with real-world situations, to demonstrate their skill and give expression to their creativity. Capital-intensive forms of production (labour-saving) are displacing human beings and rendering them more passive and

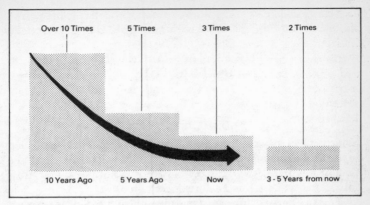

Fig. 11. The price of robots compared with the average worker's annual wage.

the systems more active. They pose serious problems for the reproduction of knowledge and the wider development of society. An attempt is made later to explore why this should be happening, and to look creatively at the alternatives which are still open to us.

Three
THE HUMAN–MACHINE INTERACTION

NONMANUAL WORK

Some readers may readily accept that this can, and has, happened in the field of manual work, but could not, or will not, occur in the field of intellectual work. They believe it is not possible to treat intellectual tasks in this Tayloristic fashion.

When a human being uses a machine, the interaction is between two dialectical opposites. The human is slow, inconsistent, unreliable but highly creative, whereas the machine is fast, reliable but totally noncreative.[1]

Originally it was held that these opposite characteristics – the creative and the noncreative – were complementary and would provide for a perfect symbiosis between human and machine, for example, in the field of computer-aided design. However, design methodology is not such that it can be separated into two disconnected elements which can then be combined at some particular point like a chemical compound. The process by which these two dialectical opposites are united by the designer to produce a new whole is a complex and as yet ill-defined and little-researched area. The sequential basis on which the elements interact is of extreme importance.

The nature of that sequential interaction, and indeed the ratio of the quantitative to the qualitative, depends on the commodity under design consideration. Even where an attempt is made to define the proportion of the work that is creative and the proportion that is noncreative, what cannot readily be stated is the point at which the creative element has to be introduced when a certain stage of the noncreative work has been completed. The process by which the designer reviews the quantitative information assembled and then makes the qualitative judgement is extremely subtle and complex. Those who seek to introduce computerised equipment into this interaction attempt to suggest that the quantitative and the qualitative can be arbitrarily divided and that the computer can handle the quantitative.

The speed at which computers are capable of carrying out immense computations is almost impossible to grasp. As long ago as the early 1960s, when the space-frame centrepiece of Expo 67 was being designed, a computer was employed for two hours. A mathematical graduate could have performed the same calculations but would have taken about 30,000 years. This is equivalent to about 1000 mathematicians working for their entire lifetimes.

THE FASTER THE BETTER

Where computer-aided design systems are installed, the operators may be subjected to work which is alienating, fragmented and of an ever increasing tempo. As the human being tries to keep pace with the rate at which the computer can handle the quantitative data in order to be able to make the qualitative value judgements, the resulting stress is enormous. Some systems we have looked at increase the decision-making rate by 1800 or 1900 per cent, and work done by Bernholz, a CAD specialist and design methodologist working in Canada, has shown that getting a designer to interact in this way will mean that the designer's creativity, or ability to deal with new problems, is reduced by 30 per cent in the first hour, by 80 per cent in the second hour, and thereafter the designer is shattered. The crude introduction of computers into the design activity, in keeping with the Western ethic 'the faster the better', may well result in the quality of design plummeting. Clearly, human beings cannot stand this pace of interaction for long.

There are arrangements in some systems where there is a set length of time for handling the data (17 seconds is an example). If you do not comply with this you are downgraded to 'head-scratching status', as the systems designers call it. The anxiety of those involved can be measured, for they display all the signs of stress such as perspiration, higher pulse rate and increased heartbeat. Suppose the image is about to disappear from the screen and you haven't finished with it. You can hold or recall it, but everyone in the office knows when you have become a head-scratcher. You are being paced by the machine, and the pace at which you work is becoming more and more visible. There comes a time when your efficiency as an operating unit is inadequate.

HUMAN 'MATERIAL'

Since employers, particularly in non academic environments, will expect the computer equipment to be used continuously, the work can be extremely stressful. In 1975, the International Labour Office recommended safeguards against the nervous fatigue of white-collar workers, and an International Federation of Information Processing working party has suggested that mental hazards 'caused by inhumanely designed computer systems should be considered a punishable offence just as endangering the bodily safety'.[2] Thus, what may be a delightfully stimulating plaything for the systems designer may be the basis for a dehumanised work environment for the user.[3]

You may think that this is an exaggeration, so let us look at what some of the leading systems designers have had to say on the subject. I quote from the American academic Robert Boguslaw and I have checked this quotation because I could not believe it could be serious when I first came across it. I was assured, however, that this statement was made after a series of discussions with some systems engineers at a major US company.

> Our immediate concern, let us remember, is the exploitation of the operating unit approach to systems design no matter what materials are used. We must take care to prevent this discussion from degenerating into the single-sided analysis of the complex characteristics of one type of systems material, namely human beings. What we need is an inventory of the manner in which human behaviour can be controlled, and a description of some of the instruments which will help us achieve that control. If this provides us with sufficient handles on human materials so that we can think of them as metal parts, electrical power or chemical reactions, then we have succeeded in placing human material on

the same footing as any other material and can begin to proceed with our problems of systems design. There are however, many disadvantages in the use of these human operating units. They are somewhat fragile, they are subject to fatigue, obsolescence, disease and even death. They are frequently stupid, unreliable and limited in memory capacity. But beyond all this, they sometimes seek to design their own circuitry. This in a material is unforgivable, and any system utilising them must devise appropriate safeguards.[4]

So according to Boguslaw, that which is most precious in human beings, the ability to design their own circuitry, or to think for themselves, is now an attribute which will quite deliberately be suppressed. The reason for all this is that the whole introduction of these systems is being based on the notion of Taylorism.

Frederick Winslow Taylor once said, 'In my system the workman is told precisely what he is to do and how he is to do it, and any improvement he makes upon the instructions given to him is fatal to success.'[5]

Taylor's philosophy is being introduced into the field of intellectual work and in order to condition us to this subordinate role to the machine and to the control of human beings through the technology, the idea is fed out in a whole series of very interesting and subtle statements. Take this one from the *Journal of Accountancy* in the United States. It talks about the idiosyncrasies of accountants and how you must control them when you introduce the computer. 'If you have got disgruntled employees, you should not allow them to start in case they might abuse the computer.' Now, I would be concerned if the computer abused the employees, but the whole philosophy is that it is the machine that matters and it is the human being who has to be modified or selected for suitability.

WHY DIMINISH THE HUMAN INTELLECT?

Professor Heath, a new-technology specialist of Heriot-Watt University, has taunted us about computers reaching an IQ of 120 'within the next two decades'. This was before 1980. He went on to say that we will then have reached the point where we shall have to

Architect or Bee?

decide whether they are people or not. I don't know what he thinks of people with an IQ of less than 120, but is this theological debate supposed to be a serious one? Professor Heath says, 'If they are people, the secular consequences are obvious. They must have the vote; switching them off would be classed as an assault and the erasure of memory as murder.'[6] Gradually we are being conditioned to think that this is a valid area of discussion.

Indeed, in Japan, over 50 per cent of industrial workers now fear the consequences of the introduction of robots for this very reason – not because they may lose their jobs but because the robots will be regarded as 'human beings' in the industrial-relations sense. In one company, the union has agreed that the robots may become union members and the company pays the subscriptions for those robot members.[7]

WHY SUPPRESS THE INTELLECT?

The more I look at human beings, the more impressed I become with the vast bands of intelligence they can use. We often say of a job, 'It's as easy as crossing a road,' yet as a technologist I am ever impressed with people's ability to do just that. They go to the edge of the pavement and work out the velocity of the cars coming in both directions by calling up a massive memory bank which will establish whether it's a mini or a bus because the size is significant. They then work out the rate of change of the image and from this assess the velocity. They do this for vehicles in both directions in order to assess the closing velocity between them. At the same time they are working out the width of the road and their own acceleration and peak velocity. When they decide they can go, they will just fit in between the vehicles.

The above computation is one of the simpler ones we do, but you should watch a skilled worker going through the diagnostic procedures of finding out what has gone wrong with an aircraft generator. There you see real intelligence at work. A human being using total information-processing capability can bring to bear synaptic connections of 10^{14}, but the most complicated robotic device with pattern-recognition capability has only about 10^3 intelligence units.

The Human–Machine Interaction

Why do we deliberately design equipment to enhance the 10^3 machine intelligence and diminish the 10^{14} intellect? Human intelligence brings with it culture, political consciousness, ideology and other aspirations. In our society these are regarded as somewhat subversive – a very good reason, then, to try and suppress them or eliminate them altogether. This is the ideological assumption present all the time. (See Figure 12.)

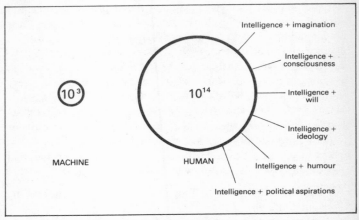

Fig. 12. Comparison of units of intelligence available for total information processing.

As designers we don't even realise we are suppressing intellects, we are so preconditioned to doing it. That is why there is a boom in certain fields of artificial intelligence. Fred Margulies, chairman of the Social Effects Committee of the International Federation of Automatic Control (IFAC), commenting recently on this waste of human brainpower, said:

> The waste is a twofold one, because we not only make no use of the resources available, we also let them perish and dwindle. Medicine has been aware of the phenomenon of atrophy for a long time. It denotes the shrinking of organs not in use, such as muscles in plaster. More recent research of social scientists

supports the hypothesis that atrophy will also apply to mental functions and abilities.

To illustrate the capabilities of human brainpower, I quote Sir William Fairbairn's definition of a millwright of 1861:

> The millwright of former days was to a great extent the sole representative of mechanical art. He was an itinerant engineer and mechanic of high reputation. He could handle the axe, the hammer and the plane with equal skill and precision; he could turn, bore or forge with the despatch of one brought up to these trades and he could set out and cut furrows of a millstone with an accuracy equal or superior to that of the miller himself. Generally, he was a fair mathematician, knew something of geometry, levelling and mensuration, and in some cases possessed a very competent knowledge of practical mathematics. He could calculate the velocities, strength and power of machines, could draw in plan and section, and could construct buildings, conduits or water courses in all forms and under all conditions required in his professional practice. He could build bridges, cut canals and perform a variety of tasks now done by civil engineers.[8]

All the intellectual work has long since been withdrawn from the millwright's function.

SQUEEZING THE MOST OUT

A decade ago, our then progressive Department of Industry produced a document that was called 'Man/Machine Systems Designing'. The different characteristics of the human being and the machine are related in this report and the different attributes listed. Under SPEED it says 'the machine is much superior' and of the human it says '1-second time lag'. Under CONSISTENCY it says of the machine 'ideal for precision' and of the human 'not reliable, should be monitored by the machine'. When it comes to OVERLOAD RELIABILITY it says of the machine 'sudden breakdown' and of the human being 'graceful degradation'.

One does not need to be a sociological Einstein to work out what is going on. The people who sell this kind of equipment make it clear enough themselves. In the *Engineer*, which I believe most engineers read, there was an advertisement for a computer-aided design package which said, 'If you've got a guy who can produce drawings non stop all day, never gets tired or ill, never strikes, is happy on half pay with a photographic memory, you don't need . . . !'[9] Now we know why that package is marketed. It states it clearly in the advert. The *Economist* likewise spells it out clearly enough. It points out 'Robots don't strike' and it advises managements to introduce robotic equipment as a way of controlling militant work forces.[10]

TOO OLD AT TWENTY-FOUR

Just as machines are becoming more and more specialised and dedicated, so is the human being, the 'appendage' to the machine. In spite of all the talk in educational circles about wider and more generalised education, the reality is that many companies will not recruit an electronics engineer over the age of twenty-three and they will specify with minute precision the exact kind of engineer and specialisation they want. The historical tendency is towards greater specialisation in spite of all the talk about universal machines and distributed systems.

The people who interface with the machine are also required to be specialised. However, as indicated above, this is accompanied by a growing rate of knowledge obsolescence. It was recently pointed out by Eugene Wigner, the internationally acclaimed physicist, when talking about the way our education system is going to meet this problem of specialisation, that it is taking longer and longer to train a physicist. 'It is taking so long to train him to deal with these problems that he is already too old to solve them.' This is at twenty-three or twenty-four years of age.

The 'peak performance age' for people of particular specialisations is being worked out by a whole range of researchers. People of different age groups sit in front of a visual display unit solving problems of growing complexity. The 'response time' is plotted against the complexity of the task as shown in the graph (Figure 13). It can be seen that as the tasks become more complex, the response time of the older people shows a much more marked

Architect or Bee?

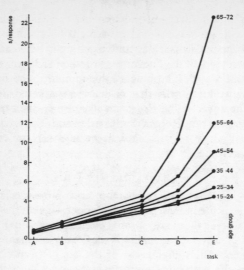

Fig. 13. Change of response time with age.

increase. So as a result of extensive scientific research, they have established that as people get older, so they get slower. Something I knew as a child of five when I looked at my grandparents!

It could be said, of course, that the older worker has a greater range of experience and knowledge and can therefore see more problems. But even if this were not the case, have we reached such a depraved stage that the natural biological process of growing old is now to be economically penalised? We design equipment to suit only the peak performance age. How many people over forty do you see in a high-pressure computerised environment, interfacing with the equipment? Yet there is nothing more natural and inevitable than growing old. 'We are all born of the gravedigger's forceps,' as Samuel Beckett once said.

The following graph (Figure 14) represents the results of experiments carried out in the USA. A group of workers (here they are scientific workers) of various ages are given some simple but

The Human–Machine Interaction

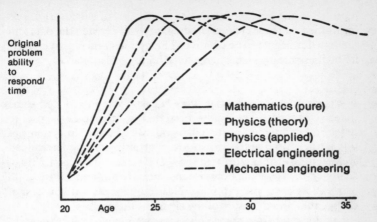

Fig. 14. Peak performance ages for various intellectual workers to achieve optimum systems interfaces.

original problems to solve. The rate at which they are solved is plotted against their age and a performance curve results.

It is found that a pure mathematician reaches his or her peak performance age at about twenty-four or twenty-five, a theoretical physicist at about twenty-six or twenty-seven and a mechanical engineer at about thirty-four. This last is the most durable profession. It happens to be my own and I am well beyond that age.

It is suggested that workers in these professions should be brought through a career pattern where they have their highest level of salary and status for a few years around their peak performance age. After that they should experience a 'career de-escalation'. This is no Orwellian projection of some grim future. It is with us here and now. Hitachi, whose chairman in Japan is seventy-three, wishes to get rid of those over thirty-five at its south Wales plant because 'older' workers are more prone to sickness, are slower, have poorer eyesight and are resistant to change.

If you have ever looked at a profile of a manual worker's pay related to physical prowess or work tempo, you will recognise that it is exactly the same kind of curve. In other words, it is now being repeated in the field of intellectual work. One of the justifications for this is that the increased productivity will provide the data and the time for people to be creative. The notion of increased creativity by these means seems to me to be highly questionable.

JUGGLE YOUR STANDARD BITS

A system known appropriately as Harness was introduced some time ago in the field of architectural design. It enables the user to reduce a building to a system of standardised units. In systems of this kind, all the architect can do is arrange the predetermined architectural elements around the VDU screen. The possibility of changing the elements becomes increasingly limited. Like a child using a Lego set, you can make pleasing patterns but you cannot change the form or nature of the elements themselves.

I understand from some colleagues who work in local government that if you use a system like Harness for about two years, you are then regarded by the architectural community as being deskilled, and have great difficulty in getting jobs. This puts the architect in a similar position to the manual worker who uses a specialised lathe and cannot then get a job doing universal and more skilled work.

Likewise, the print industry is now being transformed by the use of computers. Those in the industry are being assured that it will increase their creativity as well. Apart from the jobs permanently eliminated by these new technologies, and the conflict which arises (as at *The Times*), I would argue that much of the creative work within the print industry and the newspaper industry as a whole is being diminished. The new role of the journalists will be to work through a visual display unit where they prepare not merely the text but, through the computer, the typeface as well. It is suggested that since they can move sections of text around and modify sentences and paragraphs at great speed, their creativity will be increased. However, experience of these new technologies in the United States has already begun to show that it is resulting not in flexibility but in rigidity. This is because standard statements can be stored in the computer and called up when required to compose a story. This is done, initially, by counting, through the computer, the rate at which certain phrases or sentences occur. The most frequent ones are stored and treated as optimum sentences or 'preferred subroutines' which the journalist is then required to use. (We are obsessed with optimisation.)

Suppose you were a reporter writing about some political activity. You would have to lead in with a sentence like 'It was reported in Washington . . .' You couldn't say, for example, 'Those idiots got it wrong again' or some other unusual remark, because it would not be an available subroutine. The individual style of a journalist which gives journalism its colour and interest is gradually being diminished. There have already been complaints about some newspapers produced like this in the United States.

It is sometimes suggested that this is merely a transitional stage, a sort of industrial purgatory through which we must go on the way to a promised occupational land in which sophisticated systems and masses of data will present us with such a massive range of permutations and combinations that we can hardly fail to be highly creative. Such a view is like that of the professor with his 'contrivance' on the island of Laputa in *Gulliver's Travels*: 'Everybody knew how laborious the usual method is of attaining to arts and sciences, whereas by his contrivance, the most ignorant person, at a reasonable charge and with little bodily labour, may write books in philosophy, poetry, politics, law, mathematics and theology without the least assistance from genius or study.'

The 'contrivance' was a sort of idiot frame containing all the letters of the alphabet many times over. The pupils were trained to spin the frame continually and write down the words appearing. The logic was that if you did it often enough you could not fail to come up with something worth while – just the sort of argument used in respect of computers.

A further assumption is that 'logical' information-retrieval systems, from which we can call up dedicated packages of knowledge, will enhance our decision-making capabilities. However, as Professor Shakel of Loughborough University has pointed out, 'often human logic is not logical'.[11] Although he was talking about voice-input systems, the same may be argued for information-retrieval systems. If an intelligent human being goes to a library to look up reference material, he or she will invariably be diverted off into a series of avenues which, in terms of the dedicated knowledge required, might be regarded as redundant. Yet the richness of human behaviour and human intelligence comes about as a result of these wide bands of knowledge and experience. This apparently redundant information may subsequently be vitally important on entirely different projects and in apparently unrelated fields. We have reached a serious, if perhaps predictable, situation when *The Times* can announce in a headline, with apparent approval, 'The library where nobody browses and where automation is the chief assistant.'[12]

It is suggested that in these highly automated libraries, offices and work situations, human beings will actually enjoy conversing with machines more than with people. I have even heard it said that patients would rather converse with computers than with their doctors. This probably says more about the deplorable state of medicine in a technologically advanced society than it does for the elegance of our computer systems design. The rich interaction which comes from people discussing work problems with each other, and the open-ended intellectual cross-fertilisation which flows from that may well be lost, and human beings could become industrial Robinson Crusoes in an island of machines. This lack of human discourse and social contact, together with its effect on the functioning of the brain, has been discussed in a much wider context by the neurobiologist Steven Rose.[13]

It is typical of the narrow, fragmented and shortsighted view that our society takes of all productive processes that these important philosophical considerations are usually ignored.

LACK OF FORESIGHT

Some design methodologists have raised these questions[14] but the lack of any serious debate within the design community is itself indicative of the seriousness of the situation. One of the founders of modern cybernetics, Norbert Wiener, once cautioned, 'Although machines are theoretically subject to human criticism, such criticism may be ineffective until long after it is relevant.' It is surprising that the design community, which likes to pride itself on its ability to anticipate problems and to plan ahead, shows little sign of analysing the problems of computerisation 'until long after it is relevant'. Indeed, in this respect, the design community is displaying in its own field the same lack of social awareness which it displays when implementing technology in society at large.

Undoubtedly, most of these problems arise from the economic and social assumptions that are made when equipment of this kind is introduced. Another significant problem is the assumption that so-called scientific methods will result inevitably in better design, when in fact there are grounds for questioning whether the design process lends itself to these would-be scientific methods.[15]

Related to this is one of the unwritten assumptions of our scientific methodology – namely, that if you cannot quantify something you pretend it doesn't actually exist. The number of complex situations which lend themselves to mathematical modelling is very small indeed. We have not yet found, nor are we likely to find, a means of mathematically modelling the human mind's imagination. Perhaps one of the positive side effects of computer-aided design is that it will require us to think more fundamentally about these profound problems and to regard design as a holistic process. As Professor Lobell, the American design methodologist, has put it:

> It is true that the conscious mind cannot juggle the numbers of variables necessary for a complex design problem, but this does not mean that systematic methods are the only alternative. Design is a holistic process. It is the process of putting together complex variables whose connection is not apparent by any describable system of logic. It is precisely for that reason that the most powerful logics of the deep structures of the mind, which operate free of the limitations of space, time and causality, have traditionally been responsible for the most creative work in all of the sciences and arts. Today it has gone out of fashion to believe that these powers are in the mind.[16]

Architect or Bee?

CREATIVE MINDS

It is a fact that the highly constrained and organised intellectual environment of a computerised office is remarkably at variance with the circumstances and attributes which appear to have contributed to creativity in the arts and sciences.[17] I have heard it said that if only Beethoven had had a computer available to him for generating musical combinations, the Ninth Symphony would have been even more beautiful. But creativity is a much more subtle process. If you look historically at creative people, they have always had an open-ended, childlike curiosity. They have been highly motivated and had a sense of excitement in the work they were doing. Above all, they have possessed the ability to bring an original approach to problems. They have had, in other words, very fertile imaginations. It is our ability to use our imagination that distinguishes us from animals. As Karl Marx wrote:

> A bee puts to shame many an architect in the construction of its cells; but what distinguishes the worst of architects from the best of bees is namely this. The architect will construct in his imagination that which he will ultimately erect in reality. At the end of every labour process, we get that which existed in the consciousness of the labourer at its commencement.[18]

If we continue to design systems in the manner described earlier, we will be reducing ourselves to beelike behaviour.

It may be regarded as romantic or succumbing to mysticism to emphasise the importance of imagination and of working in a non linear way. It is usually accepted that this type of creative approach is required in music, literature and art. It is less well recognised that this is equally important in the field of science, even in the so-called harder sciences like mathematics and physics. Those who were creative recognised this themselves. Isaac Newton said, 'I seem to have been only like a boy playing on the sea shore and diverting myself in now and then finding a smoother pebble or a prettier shell than ordinary, while the great ocean of truth lay all undiscovered before me.'

Einstein said, 'Imagination is far more important than knowledge.' He went on to say, 'The mere formulation of a problem is far more important than its solution which may be merely a matter of mathematical or experimental skills. To raise new questions, new possibilities and to regard old problems from a new angle requires creative imagination and marks real advances in science.'

On one occasion, when being pressed to say how he had arrived at the idea of relativity, he is supposed to have said, 'When I was a child of fourteen I asked myself what the world would look like if I rode on a beam of light.' A beautiful conceptual basis for all his subsequent mathematical work.

Central to the Western scientific methodology is the notion of predictability, repeatability and quantifiability. If something is unquantifiable we have to rarefy it away from reality, which leads to a dangerous level of abstraction, rather like a microscopic Heisenberg principle. Such techniques may be acceptable in narrow mathematical problems, but where much more complex considerations are involved, as in the field of design, they may give rise to questionable results.

> The risk that such results may occur is inherent in the scientific method which must abstract common features away from concrete reality in order to achieve clarity and systematisation of thought. However, within the domain of science itself, no adverse results arise because the concepts, ideas and principles are all interrelated in a carefully structured matrix of mutually supporting definitions and interpretations of experimental observation. The trouble starts when the same method is applied to situations where the number and complexity of factors is so great that you cannot abstract without doing some damage, and without getting an erroneous result.[19]

More recently, these questions have given rise to a serious political debate on the question of the neutrality of science and technology[20] and there is likely to be growing concern over the ideological assumptions built into our scientific methodologies.[21]

Four
COMPETENCE, SKILL AND 'TRAINING'

THE ORIGINS OF DESIGN

Around the sixteenth century, there appeared in most European languages the term 'design' or its equivalent. The emergence of the word coincided with the need to describe the occupational activity of designing. That is not to suggest that designing was a new activity. Rather, it was separated out from a wider productive activity and recognised as an activity in its own right. This can be said to constitute a separation of hand and brain, of manual and intellectual work and of the conceptual part of work from the labour process. Above all, it indicated that designing was to be separated from doing.

It is clearly difficult to locate a precise historical turning point at which this occurred; rather we will view it as a historical tendency.

Up to the stage in question, a great structure such as a church would be 'built' by a master builder. We may generalise and say that the conceptual part of work would be integral to that labour process. Thereafter, however, came the concept of 'designing the church', an activity undertaken by architects, and the 'building of the church', an activity undertaken by builders. In no way did this represent a sudden historical discontinuity, but it was rather the beginning of a discernible historical tendency which has still not worked its way through many of the craft skills, so that as recently as the last century, Fairbairn was able to give his comprehensive description of the skills of a millwright quoted in Chapter 3. To this day, there are many jobs in which the conceptual part of work is still integrated with the craft basis. The significant feature of the stage in question is, however, that separating manual and intellectual work provided the basis for further subdivisions in the field of intellectual work itself – or, as Braverman put it, 'Mental labour is first separated from the manual labour and then itself is subdivided rigorously according to the same rules'.[1]

Dreyfus locates the root of the problem in the Greek use of logic and geometry, and the notion that all reasoning can be reduced to some kind of calculation. He suggests that the start of artificial intelligence probably began around the year 450 BC with Socrates and his concern to establish a moral standard. He asserts that Plato generalised this demand into an epistemological demand where one might hold that all knowledge could be stated in explicit definitions which anybody could apply. If one could not state one's know-how in such explicit instructions, then that know-how was not knowledge at all but mere belief. He suggests a Platonic tradition in which, for example, cooks, who proceed by taste and intuition, and people who work from inspiration, like poets, have no knowledge. What they do does not involve understanding and cannot be understood. More generally, what cannot be stated explicitly in precise instructions – that is, all areas of human thought which require skill, intuition or a sense of tradition – is relegated to some kind of arbitrary fumbling.[2]

Gradually, a view evolved which put the objective above the subjective, and the quantitative above the qualitative. That the two should and can interact was not accepted, in spite of a systematic effort and intellectual struggle to assert it. One important example of the attempt to do so was the work of Albrecht Dürer (1471–1528). Dürer was not only a 'Master of the Arts' but a brilliant mathematician as well, who reached the highest academic levels in Nuremberg. Dürer sought to use his abilities to develop mathematical forms which would succeed in preserving the unity of hand and brain. Kantor[3] points out the significance of Dürer's ability to put complex mathematical techniques to practical uses, while Olschki[4] compares his mathematical achievements with those of the leading mathematicians in Italy and elsewhere at that time. Indeed, some ninety years after Dürer's death Kepler was still discussing his geometric construction techniques. Alfred Sohn Rethel points out, in speaking of Dürer, 'Instead of, however, using this knowledge in a scholarly form, he endeavoured to put it to the advantage of the craftsman. His work was dedicated "to the young workers and all those with no one to instruct them truthfully". What is novel in his method is that he seeks to combine the workman's practice with Euclidian Geometry'. And further:

> What Dürer had in mind is plain to see. The builders, metalworkers, etc., should on the one hand be enabled to master the tasks of military and civil technology and of architecture which far exceeded their traditional training. On the other hand the required mathematics should serve them as a means, so to speak, of preserving the unity of head and hand. They should benefit from the indispensable advantages of mathematics without becoming mathematic or brainworkers themselves. They should practise socialised thinking yet remain individual producers, and so he offered them an artisan schooling in draughtsmanship permeated through and through with mathematics – not to be confused in any way with applied mathematics.[5]

It was said that on one occasion Dürer proclaimed it would be possible to develop forms of mathematics that would be as amenable to the human spirit as natural language. Thereby one could integrate into the use of the instruments of labour the conceptual parts of work, thus building on the tradition in which the profiles of complex shapes were defined and constructed with such devices as sine bars.

HOLISTIC DESIGN

Thus theory, itself a generalisation of practice, could have been reintegrated into practice to extend the richness of that practice and application while retaining the integration of hand and brain.

The richness of that practical tradition may be found in the sketchbook of Villard de Honnecourt, in which he introduced himself thus:

> Villard de Honnecourt greets you and begs all who will use the devices found in this book, to pray for his soul and remember him. For in this book will be found sound advice on the virtues of masonry and the uses of carpentry. You will also find strong help in drawing figures according to the lessons taught by the art of geometry.[6]

Competence, Skill and 'Training'

This extraordinary document by a true thirteenth-century cathedral builder contains subjects which might be categorised as follows:
1. Mechanics
2. Practical Geometry and Trigonometry
3. Carpentry
4. Architectural Design
5. Ornamental Design
6. Figure Design
7. Furniture Design
8. 'Subjects foreign to the special knowledge of Architects and Designers'

The astonishing breadth and holistic nature of the skills and knowledge are in the manuscripts for all to see.

There are those who, while admitting to the extraordinary range of capabilities of craftspeople of this time, hold that it was a 'static' form of knowledge which tended to be handed unaltered from master to apprentice. In reality, there was in these crafts and in their transmission embodied dynamic processes for extending their base and adding new knowledge all the time. Some of the German manuscripts describe *die Wanderjahre* – a form of sabbatical in which craftspeople travel from city to city to acquire new knowledge. Villard de Honnecourt travelled extensively, and thanks to his sketchbook we can trace his travels through France, Switzerland, Germany and Hungary.

He was also passionately interested in mechanical devices, and one system he designed was subsequently adapted to keep mariners' compasses horizontal and barometers vertical. He devised a variety of clock mechanisms, from which we learn 'how to make the Angel keep pointing his finger towards the sun', and he displayed extraordinary engineering skills in a range of lifting and other mechanisms to provide significant mechanical advantage. For example, he invented a screw combined with a lever with appropriate instructions – 'How to make the most powerful engine for lifting weights.'

In all this, we see brilliantly portrayed the integration of design with doing – a tradition which was still discernible when Fairbairn described his millwright.

Architect or Bee?

Villard was also concerned with 'automation', but in a form which freed the human being from backbreaking physical effort but retained the skilled base of work. In woodworking, he thought of a system of replacing the strenuous sawing activity – 'How to make a saw operate itself.'

He was profoundly interested in geometry as applied to drawings: 'Here begins the method of drawing as taught by the art of geometry but to understand them one must be careful to learn the particular use of each. All these devices are extracted from geometry.' He proceeds to describe 'How to measure height of a tower', 'How to measure the width of a water course without crossing it', 'How to make two vessels so that one holds twice as much as the other'.

Many modern researchers have testified to Villard's significant grasp of geometry. Side by side with this we find his practical advice to stonecutters on the building elements dimensions: 'How to cut an oblique voussoir', 'How to cut the springing stone of an arch', 'How to make regular pendants'. All of this latter, drawn from his own practical experience and skill, is a vivid portrayal of the integration of hand and brain.

Another thirteenth-century manuscript, written in the same dialect as Villard's, is still preserved and can be consulted in the Bibliothèque St Geneviève in Paris. Its author likewise concerned himself with mathematical problems: 'If you want to find the area of an equilateral triangle', 'If you want to know the area of an octagon', 'If you want to find the number of houses in a circular city'.

Throughout this period, the intellectual and the manual, the theory and the practice were integral to the craft or profession. Indeed, so naturally did the two coexist that we find practical builders (architects) with the university title like *Doctor Lathomorum*.

The epitaph of Pierre de Montreuil, the architect who reconstructed the nave and transepts of Saint Denis, runs, 'Here lies Pierre de Montreuil, a perfect flower of good manners, in this life a Doctor of Stones.'

I have cited these sketchbooks and quoted from these manuscripts in order to demonstrate that the craft at that time embodied powerful elements of theory, scientific method and the conceptual or design base of the activity. In doing so, I am myself guilty of a serious error. I accept that a matter can only be scientific or theoretical when it is written down. I did not provide an illustration of a great church or complex structure and state that the building of such a structure must itself embody a sound theoretical basis, otherwise the structure could not have been built in the first instance.

We can also detect in the written form the basic elements of Western scientific methodology: predictability, repeatability and mathematical quantifiability. These, by definition, tend to preclude intuition, subjective judgement and tacit knowledge.

Furthermore, we begin to regard design as something that reduces or eliminates uncertainty, and since human judgement, as distinct from calculation, is itself held to constitute an uncertainty, it follows some kind of Jesuitical logic that good design is about eliminating human judgement and intuition. Furthermore, by rendering explicit the 'secrets' of craft, we prepare the basis for a rule-based system.

'RULES' FOR DESIGN

In the two successive centuries there followed systematic attempts to describe and thereby render visible the rules underlying various craft skills. This applied right across the spectrum of skills of people who were artists, architects and engineers, in the Giotto tradition, from the theory of building construction through to painting and drawing. Giotto's method was not precisely optical. The receding beams of the ceiling converge to a reasonably convincing focus, but it is only approximate and does not coincide with the horizontal line as it should, according to the rules of linear perspective. 'This method is, however, systematic and rational, factors which no doubt provided a powerful stimulus for the more fully scientific rule seekers of the subsequent centuries. Priority amongst those who preceded Leonardo in searching for precise optical laws in picture making must go to the great architect and erstwhile sculptor Filippo Brunelleschi.'[7]

According to Manetti, at some time before 1413 Brunelleschi

constructed two drawings which showed how buildings could be represented 'in what painters today call perspective, for it is part of that science which in effect sets down well and with reason the diminutions and enlargements which appear to the eyes of man from things far away and close at hand'. One of the paintings showed the octagonal baptistry (S. Giovanni) as seen from the door of the cathedral in Florence. The optical 'truth' was verified by drilling a small hole in the baptistry panel, so that the spectator could pick up the panel and press an eye to the hole on the unpainted side and, with the other hand, hold a mirror in such a way that the painted surface was visible in reflection through the hole. By these means, Brunelleschi established precisely the perpendicular axis along which his representation should be viewed.

By the use of a mirror, there was a precise matching of the visual experience and the painted representation, and this was to become Leonardo's theory of art and indeed his whole theory of knowledge.[8] He applied the same scientific methods to his architectural and other designs. One interpretation of these events is that they represented a significant turning point in the history of design and design methodology. Thereafter, there is a growing separation of theory and practice, a growing emphasis on the written 'theoretical forms of knowledge' and, in my view, a growing confusion in Western society between linguistic ability and intelligence (in which the former is taken to represent the latter). Furthermore, this is accompanied by a growing denigration of tacit knowledge in which there are 'things we know but cannot tell'.[9] We may cite that most illustrious embodiment of theory and practice – Leonardo da Vinci: 'They will say that not having learning, I will not properly speak of that which I wish to elucidate. But do they not know that my subjects are to be better illustrated from experience than by yet more words? Experience, which has been the mistress of all those who wrote well and thus, as mistress, I will cite her in all cases.'[10]

In spite of such assertions, the tendency to produce generalised, written-down, scientific or rule-based design systems continued to build on earlier work. In 1486, the German architect Mathias Roriczer published in Regensburg his deceptively named 'On the

Ordination of Pinnacles'. In this, he set out the method of designing pinnacles from plan drawings, and in fact produced a generalised method of design for pinnacles and other parts of a cathedral. These tendencies had already elicited bitter resistance from the craftsmen-cum-designers whose work was thereby being deskilled.

THE MASTER MASONS

In 1459, master masons from cities like Strasburg, Vienna and Salzburg met at Regensburg in order to codify their lodge statutes. Among the various decisions, they decided that nothing was to be revealed of the art of making an elevation from a plan drawing to those who were not in the guild. 'Therefore, no worker, no master, no wage earner or no journeyman will divulge to anyone who is not of our Guild and who has never worked as a mason, how to make the elevation from the plan.' Of particular note is the exclusion of those who had never *worked* as a mason.

There is, as our German colleagues would put it, a *Doppelnatur* to this craft reaction. On the one hand there is the negative elitist attempt to retain privileges of the profession rather as the medical profession seeks to do to this day. On the other, there is a highly positive aspect, that of attempting to retain the qualitative and the quantitative elements of work, the subjective and the objective, the creative and the noncreative, the manual and the intellectual, and the work of hand and brain, embodied in the one craft.

The pressures on the master masons were twofold. Not only was the conceptual part of the work to be taken away from them, but those workers who still embodied the intellectual and design skills were being rejected by those who sought to show that theory was above, and separate from, practice. The growing academic elite resented the fact that carpenters and builders were known as masters, for example, *Magister Cementarius* or *Magister Lathomorum*. The academics attempted to ensure that '*Magister*' would be reserved for those who had completed the study of the liberal arts. Indeed, as early as the thirteenth century, doctors of law were moved to protest formally at these academic titles for practical people.

It would be both fascinating and illuminating to trace these tendencies through the five intervening centuries which take us up to the information society of computer-aided design and expert systems, but space does not permit it in this book. Suffice it to say that a number of researchers, drawing on historical perspective and viewing the implications of these information-based systems, conclude that we may now be at another historical turning point where we are about to repeat, in the field of design and other forms of intellectual work, many of the mistakes made in the field of craftsmanship in the past.[11, 12]

SEPARATION OF THEORY FROM PRACTICE

It is significant that J. Weizenbaum, a professor of computer science at the Massachusetts Institute of Technology, uses the subtitle 'from judgement to calculation' in his seminal work *Computer Power and Human Reason*[13] and highlights the dangers which will surround an uncritical acceptance of computerised techniques.

The spectrum of problems associated with them is already becoming manifest. They include such spectacular separation of theory and practice as to result in some of those who have been weaned on computer-aided design being unable to recognise the object that they have 'designed'. Epitomising this was the designer of an afterburner igniter who calculated the dimensions on the CAD screen and then set them out with the decimal point one place to the right (which, in an abstraction, is very much like one place to the left). He then generated the numerical control tapes with which deskilled workers on the shop floor produced an igniter ten times larger than it should have been.[14] Perhaps the most alarming aspect of this extraordinary state of affairs was that when confronted with the monstrosity, the designer saw nothing wrong with it.

Less spectacular, but in the long term of growing significance, is the design rigidity which menu-driven systems tend to produce. Harness, described in Chapter 3, is an example.

Given the scale and nature of these problems and the exponential rate of technological change within which they are located, it behoves all of us to seek to demonstrate, as Dürer did, that alternatives exist which reject neither human judgement, tacit knowledge, intuition and imagination nor the scientific or rule-

Competence, Skill and 'Training'

based method. We should rather unite them in a symbiotic totality.

Unfortunately, there are few examples of such 'symbiotic' systems, that is, systems where the pattern-recognition abilities of the human mind, its assessment of complicated situations and intuitive leaps to new solutions are combined with the numerical computation power of the computer. They do exist in narrow specific areas, as Professor Howard Rosenbrock of the Control Systems Group at UMIST has demonstrated with the computer-aided design of complex control systems where the performance is displayed as an inverse Nyquist array on the screen.[15] I have myself described the potential for human-centred systems both in skilled manual work and in design.[16] Furthermore, in the technology division of the Greater London Enterprise Board, we have been working on the development of expert medical systems through our technology networks. (See Chapter 8.) These provide an interaction between the 'facts of the domain' and the fuzzy reasoning, tacit knowledge, imagination and heuristics of the expert, and no attempt is made to reduce all these aspects to a rule-based system – the system is seen as something that aids rather than replaces the expert.

An important breakthrough for these human-centred systems has been the recent decision by the EEC's ESPRIT programme to fund jointly a project to build the world's first human-centred computer-integrated manufacturing system.[17] Details of this are given in Chapter 8. Professor Rauner, Professor Wittowsky and their colleagues at the University of Bremen are developing an educational program which will go with the system since we are concerned not merely with production but with the reproduction of knowledge. Those working on the educational package are practical engineers themselves.

CONSUMER INCOMPETENCE

Efforts to deskill the producers can only become operational if they are accompanied by the deskilling of the consumers. The deskilling of bakers, for example, can come about only if that awful cotton-wool stodge in plastic packets is regarded as bread by millions of consumers. Highly automated and factory farming techniques are only possible if the public believes that there are only two kinds of potato, 'new' and 'old', that cookers and eaters are the only forms of apple, and if it cannot distinguish the taste of free-range poultry and eggs from those produced under battery-farm conditions.

The elimination of high-level skills in carpentry and cabinet-making is possible because large sections of the public do not appreciate the difference between a tacky chipboard product and one handmade with real wood and fitted joints, or between a plastic container and (say) an inlaid needlebox.

The concern for quality should not be misunderstood as an elitist tendency. Quite ordinary working-class and rural families used to pass pieces of furniture from one generation to another which, although simple, embodied fine craftsmanship and materials. A skilled joiner recently told me with great feeling how monstrous he found it that beautiful pieces of wood which could have been hand-turned and carved were being burned on a demolition site by 'builders' who couldn't distinguish between one piece of wood and another.

Given time, more and more sections of the community will lose the capacity to appreciate craftsmanship and goods of quality. As you 'break the refractory hand of labour', you must also break the refractory will of the consumer. To do so it is of course necessary to have ranges of accomplices. These are in advertising, marketing and, more generally, of the *Waste Makers* type. These accomplices are in relation to production and consumption, and there are also partners in crime in the areas of the reproduction of knowledge. The duality of the master and apprentice, teacher and student, has now been replaced by the trainer and the trainee. In large occupational areas, we no longer have education, we have 'training'.

APPRENTICESHIPS AND 'TRAINING'

An apprenticeship in the classical sense was not merely a process for the acquisition of technical skills. It was far more significantly the transmission of a culture, a way of understanding and respecting quality and acquiring a love of good materials. Even to this day, this cultural outlook is alive and well among craftspeople.

Ken Hunt is a master engraver whose work is sought worldwide. He served his apprenticeship with Purdeys, the London Sporting

Gunmakers, who arranged for him to work with Henry Kell, one of the specialist firms engraving the gun actions. This is how Ken Hunt described his work in 1987:

> I think engraving creates an intensely personal relationship between the work and the craftsman. It's the most lovely feeling when everything is going right; the cutting tool is working well, the steel doesn't fight you.
>
> To me, the beauty of a cut on steel with a graver is similar to the mark made by a quill pen on paper. It flows and tapers and is far removed from the straight line drawn by a ball point. I get so involved sometimes that I lose all track of time, and I get lost in all sorts of ideas, almost fantasies, I suppose. I find myself thinking of craftsmen centuries ago who worked metal in exactly the same way as I do now. Nothing has changed, neither the medium nor the tools.
>
> It may sound strange, but occasionally I get pieces back that I might have worked on in the sixties or even earlier, and I only have to touch them to recall exactly what I was doing and thinking when I was working on them all those years ago. Perhaps it's because each job represents and absorbs a large part of your life – maybe even your soul, who knows? Michaelangelo used to claim that all he did when confronted with a block of stone was to chip away and release the sculpture which was inside it, and I feel that too.

Ken does not use preliminary drawings of his intended work. Nor does he have tracing on the metal which will then be simply followed by the engraving tool. 'No, I go straight in and just do it. I've got an idea in my head as to how the finished work will look, but I don't believe in drawing it out carefully first.'

There is a tendency to regard such craft skills as being static and devoid of development. But the environment created by an apprenticeship encourages experimentation and innovation within a given tradition. Ken Hunt recalls that in his early days he would visit museums to admire and wonder at masterpieces from the past.

> I would stand and stare at a certain piece for ages wondering how it was done. Sometimes, I would even stay so long that the

wardens would begin to eye me with suspicion! I was intrigued with everything to do with metal work, though especially gold inlaying. I eventually worked out my own way of keying gold to steel using a series of undercut crisscross lines which have a dovetail effect.[18]

It would be unthinkable that craftspeople like Ken Hunt would waste materials or mishandle or damage tools and equipment. *All* of this was integral to the totality which was embodied in a traditional apprenticeship. It was also a process by which one learned, in a very practical way, the logistics of procuring such materials, treating them and forming them in a creative process which linked hand, eye and brain in a meaningful productive process. It embodied 'design by doing' – methods of work in which the conceptual aspects of work were integrated within the overall labour process.

Apprenticeships served to develop significant skills in the field of planning and coordination, and produced quite astonishing levels of ability in the handling of materials. I marvel at St Paul's Cathedral, for even given our modern means of project management and complex techniques for handling material, we may question whether anyone would be capable of constructing it today. Even if we could, what an infinitely greater task it was in the seventeenth century, given the limited equipment for lifting and placing the building elements into their locations.

The kind of apprenticeships those builders had gave them a deep sense of total machines as operating systems, epitomised by the vast knowledge of the great millwrights (see Chapter 3). It is true that with the introduction of Taylorism,[19] apprenticeships did embody almost anecdotal aspects, where considerable time was spent in making tea for others or in irrelevant activities, but that is not what we are addressing here; it is rather the great apprenticeships which produced those of the calibre described earlier in the chapter.

Against this richness and competence can be counterposed 'training'. The word is very apt in the modern context. My own hierarchy of verbs in terms of competence transmission would be

Competence, Skill and 'Training'

the following: you program a robot, you train a dog (or possibly a soldier), but for human beings you provide educational environments. Training produces narrow, overdedicated capabilities which are generally machine, system or program-specific. With the ever increasing rate of technological change, the 'knowledge' required to cope with a particular machine or system may be obsolete in a couple of years' time. The trainee is then lost, and requires further 'training'. Much of what now passes for 'training' is nothing more than a form of social therapy. Instead of putting people on Valium you put them on a training course. It is questionable whether you produce anything more than a slightly better-quality dole queue.

'Training' often hides a cruel deception. In some of the inner-city areas single-parent women on training courses are led to believe that if they can fiddle around with a BBC micro, they are then information technologists, and the multinationals will be beating the way to their door to offer them work. This complete misunderstanding of the levels of skill required for given activities in the real world, and the manner in which those skills are acquired – particularly diagnostic skills – is outside the range of this new moribund layer of 'trainers'.

Some companies have very competent training officers who themselves have actual knowledge of the processes involved. What I am referring to here is that new band of 'training advisers', 'training coordinators', 'training outreach workers' and 'training planners' who seem to believe that there is some separate activity called 'training' which transcends all other forms of professional knowledge. Some of the ones I have encountered seem to believe that if you've trained a Labrador to retrieve you can also train a nuclear physicist, and if you've trained somebody to make doughnuts in the catering industry, you can also train them to design a Rolls-Royce aero engine; it is, after all, just training! Because these people have no knowledge of the skills involved, they behave in a high-handed and arrogant fashion. Furthermore, because they are in a position to allocate funds, they are often able to impose their nonsense on people who could have provided a rich developmental environment.

Architect or Bee?

The disadvantage of using this type of 'trainer' is twofold. They don't know what they are doing and are overpaid for doing it, and, more significantly, they prevent people who do have the skills and knowledge from enjoying the experience and gaining the dignity of transmitting it to a future generation. Some of the 'quality training schemes' set up by the progressive local governments are particularly hideous in this regard.

'TRAINING' AND DESTRUCTION OF SKILL

I have seen so-called training advisers and coordinators who wouldn't know one end of a building from another, travelling out in chauffeur-driven cars to building sites to decide whether they would let skilled building workers have trainees with them, and questioning them on whether they had been on this little course or that little course organised by 'trainers' like themselves. The fact that these workers had been passing their skills on to apprentices for years and that these apprentices had developed into people who produced real structures rather than long, boring, irrelevant reports was something they did not consider important. They even insisted on having, for a pathetic little 'taster' course in workshop practice, a massive report prepared which included many pages listing the tools the trainee would learn to handle. There were pages which stated 'The trainee shall be acquainted with the use of a flat file, a square file, a round file, a half round file, a triangular file . . .' and this continued until all the conceivable file forms one might encounter were listed. Every tool set likewise had to be listed. All of this was to prepare more reports, so that Filofax socialists, now calling themselves trainers, could satisfy themselves, by the sheer bulk of the report, that they had created a 'quality scheme'. All this was necessary because they hadn't the slightest idea what was going on. It really was a scorching example of 'Never mind the quality, feel the width'. A skilled person would simply have stated that the young person was capable of 'handling the tools of the trade'.

One 'training adviser' actually assured me that she had 'designed a building course' to produce a builder in one year. This was all the more extraordinary since she had never been near the building industry, had no idea of its skills, requirements or practices. It transpired that she was not talking about a builder at all, but rather somebody who could do a bit of bricklaying. The idea that a builder could construct a complete house, as skilled builders have done in the past – or can still do in many rural areas – was totally outside this person's range of experience or expectation. It is precisely because of these systems-trained so-called builders that we in Britain have had many of the failures in building construction which have resulted in miserable living conditions, accidents and, more recently, decisions to demolish and start again. New courses like this constitute a destruction of craft skills.

There is too the notion that such courses should be 'scientific'. These so-called scientific methods are held to be infinitely more important than the process of learning by doing or by working with and gaining knowledge from people who know what they are doing. In one case a group of skilled workers had to undergo a course on 'project management'. This had to do with the description of the project, the planning of the project environment, the preparation of the site, the procurement of the materials, the special tools required, the sequential steps to be taken and the assessment of the project outcome. The actual project or 'case study' which was subjected to this level of abstraction was 'repairing a tap' (changing a washer in one).

The lecture went on for nearly two hours of waffle. One of the course members, a ships engineer who had repaired high- and low-pressure, hydraulically and pneumatically operated valves, said he was so confused by all this 'theory' that he began to doubt whether in fact he would be capable of repairing a tap at all. He would, he said, have great difficulty in writing the long and boring account of how it was done. The fact that he could change a washer in a few seconds was regarded as quite irrelevant in becoming a qualified trainer by the lecturer, whose own background was in catering.

The whole attitude behind this form of training also shows that even 'progressive' local governments do put administrators and bureaucrats into positions of power and authority over those who can actually do things. When they talk about discrimination and equal opportunities, they certainly do very little to provide equal opportunities for manual workers and those whose knowledge of the world is experiential and real. I know of cases where after an interview for a job as a trainer a skilled craftsperson who had come from precisely the craft and industry involved and had answered every question absolutely correctly, if very briefly, was judged by the panel as having 'little to say'. When subsequently they interviewed a sociologist who hadn't the slightest idea of the skills involved, having never been near the industry, but could waffle on about theories of knowledge acquisition, the panel were most impressed. The sociologist got the job, and made a complete mess of it. This once again underlines the deep confusion between linguistic ability and competence.

Five
THE POTENTIAL AND THE REALITY

THE POTENTIAL

Those who initiate scientific and technological advances are frequently moved by the loftiest motives and display a genuine desire to improve the quality of life of those affected by their innovations. Who would doubt the motives of Pascal who, when he had designed and built the first true mechanical calculating machine in 1642, declared, 'I submit to the public a small machine of my own invention, by means of which you alone may, without any effort, perform all the operations of arithmetic, and may be relieved of the work which has so often fatigued your spirit when you worked with the counters and with the pen.'

The motives of those innovating in the field of computer-aided design are likely to be equally laudable. Professor Tom Mayer and his colleagues at Strathclyde would like to see computers used to democratise the decision-making process in architectural design. Arthur Llewelyn, a leading CAD specialist in the UK, has repeatedly asserted that computers should not be used as a means of eliminating designers and draughtsmen, but rather as tools to improve their responsibility and ability to carry out creative tasks.

Regrettably, the history of scientific and technological innovation is strewn with dramatic examples which contrast the dedicated and socially desirable objectives of the academic or researcher with the cynical exploitation of their ingenuity at the level of application by the owners of the means of production. Hence we find, in many fields of endeavour, a significant gap between what technology could provide (its potential) and what it does provide (its reality).

There is a tendency, therefore, to make value judgements about given technologies based on what they might achieve rather than what they have already achieved and are likely to continue to achieve within the given economic, political and social framework.

Thus, those who have no experience of CAD (or are still at the gee-whiz stage) tend to display a more positive attitude to CAD than those who have had to live with it for some time. Similarly, the genuine enthusiasm of a CAD specialist on a research project in the relative monastic quiet of a university is unlikely to be shared by the designer faced with the harsh reality of its consequences in some high-pressure multinational corporation.

In academic circles, concern has recently been expressed that if we do not properly understand this historical conjunction, we may well pursue a technological course which will permanently close off options for more humane and satisfying organisational forms in the field of intellectual work, in much the same way as we have already done in the field of craftsmanship. Failure to recognise that these options are still open to us in CAD, and that we still have the time and indeed the responsibility to question the linear drive forward of this technology, may well mean that we shall see growing alienation and loss of job satisfaction in engineering design. This is likely to be accompanied by the subordination of the operator (designer) to the machine (computer), with the narrow specialisation of Taylorism leading to the fragmentation of design skills and a loss of the panoramic view of the design activity itself. In consequence, standard routines and optimisation techniques may seriously limit the creativity of the designer, because the subjective value judgements would be dominated by the 'objective' decision of the system. To put it another way, the quantitative elements of the design activity will be regarded as more important than the qualitative ones. There is already evidence to show that CAD, when introduced on the basis of so-called efficiency, gives rise to a deskilling of the design function and a loss of job security – particularly for older men giving way to structural unemployment.

To analyse why these contradictions should arise, it seems necessary to view the computer as part of a technological continuum, and its consequences as those that arrive when high-capital equipment is introduced into any work environment, whether it be manual or intellectual. It must also be analysed within the economic, social and political context of the society which has given rise to the technology itself.

The Potential and the Reality

INDICATORS

If a comparison between design (intellectual work) and skilled craftsmanship (manual work) is really tenable, we will increasingly find strong indicators of the following:

a. The subordination of the operator (designer) to the requirements of the machine (computer) with shift work or systematic overtime to counter the increasing rate of obsolescence of the machine.
b. Emphasis on machine-centred systems rather than human-centred ones.
c. Limitation of the creativity of the designer by standard routines and optimisation.
d. Domination of the subjective value judgements of the designer by the 'objective' decisions of the system. That is, the quantitative elements of design will be treated as more important than the qualitative ones.
e. Alienation of the designer from his or her work.
f. Abstraction of the design activity from the real world.
g. A fragmentation of design skills (overspecialisation) with a loss of panoramic view, together with the introduction of Taylorism and other forms of 'scientific management', even to the extent of measuring the rate of performing intellectual work.
h. Deskilling the design function.
i. Increased work tempo as the designer is paced by the computer.
j. Increased stress, both physical and mental.
k. Loss of control over one's work environment.
l. Growing job insecurity, particularly for older men.
m. Knowledge obsolescence.
n. The gradual proletarianisation of the design community as a result of the tendencies indicated above, and, in consequence of this, the considerable increase in trade-union membership and industrial militancy.

THE REALITY

CAD equipment shares with all high-capital equipment in a profit-oriented society the contradiction of an increasing obsolescence rate (the increasingly short life of fixed capital). Sophisticated CAD equipment is now obsolete in about three years. In addition, the investment cost of the means of production (as distinct from the price of individual commodities) is ever increasing. As owners of equipment which is becoming obsolete literally by the minute, and which has required enormous capital investment, employers will seek to exploit it twenty-four hours a day. This trend has long been evident on the shop floor and the effects of shift working are already well documented. The same problems are beginning to be quite evident in the field of white-collar work.[1]

As far back as the early 1970s the AUEW-TASS (Amalgamated Union of Engineering Workers, Technical, Administrative and Supervisory Section) was in a major dispute with Rolls-Royce which cost the union £250,000. The company sought, among other things, to impose on the design staff at its Bristol plant the following conditions:

1. The acceptance of shift work in order to exploit high-capital equipment.
2. The acceptance of work-measurement techniques.
3. The division of work into basic elements and the setting of times for these elements, such times to be compared with actual performance.

In this particular case, industrial action prevented the company from imposing these conditions. They are, however, the sort of conditions that employers will increasingly seek to impose on their white-collar workers.

When staff workers, whether they be technical, administrative or clerical, work in a highly synchronised computerised environment, the employer will try to ensure that each element of their work is ready to feed into the process at the precise time at which it is required. A mathematician, for example, will find that he has to have calculations ready in the same way as a Ford worker has to have the wheel ready for the car as it passes him on the production line. Consequently we can say that the more technological change and computerisation enter into white-collar areas, the more workers in these areas will become proletarianised. The consequences of shift work will spread across the family, social and cultural life of the white-collar worker.

In a survey carried out in West Germany,[2] it was demonstrated that the ulcer rate of workers on a rotating shift was eight times higher than that of other workers.

A higher proportion of night and rotating shift workers reported that they were fatigued much of the time, that their appetites were dulled and that they were often constipated.

The most frequently mentioned difficulties in husband/wife relationships concerned the absence of the worker from the home in the evenings, sexual relations and difficulties encountered by the wife in carrying out her household duties.

Another area of family life that seems to be adversely affected by certain kinds of shift work is the parent/child relationship.

I quote these extracts without making any judgement about the nuclear family. I am simply indicating that the nature of technology produces effects which spread right through the fabric of society to affect the way we live and the way we relate to other people.

The disruption of social life outside the family is also considerable. I was at one time acquainted with a suburban estate in west London where a number of mathematics graduates worked. They used to participate in activities like badminton, local operatics and a theatre group. When the large firm in which some of them worked introduced a computerised system, it required them to work on shift. Consequently their other activities were completely disrupted.

Thus, in practice, there are grounds already for suggesting that in white-collar work, far from humanising the nature of it, high-capital equipment is diminishing the quality of life of intellectual workers just as it has already done to shop-floor workers.[3]

VDUs

The tendency towards automation in offices leads to a reduction in the volume of paper, and to systems which are of 'high information density'. Micrographics systems are now commonplace peripherals to computerised systems. Complaints of eyestrain, visual discomfort, difficulties in reading and postural fatigue are now widespread. Östberg has described some of these difficulties in an important paper which contains eighty-four literature references.[4]

The effects of ageing are significant for the users of these micrographic systems. Typically, for a person sixteen years old with normal eyesight, about 12 dioptres of accommodation are available (the near point being 8 cm) of which only one dioptre (near point at 100 cm) remains at sixty. In consequence, employees over the age of fifty are frequently regarded as 'visually handicapped' and unsuitable for long term work with these systems. Increasingly (particularly in Sweden), trade-union and health and safety representatives are demanding that such systems should be designed to accommodate a wide age spectrum. These demands are part of the growing international insistence by workers on the right to be involved in the design of their jobs, work stations and wider working environment.

By 1980 there were between 5 and 10 million VDUs in use. They have since become commonplace in every office of significant size. Even so, the controversy still continues about the effects on the user – in particular of the low-level radiation emitted. This controversy has been going on for the past twenty years. In 1968, a meeting of over 100 European experts reached the conclusion that working with VDUs for eight hours causes fatigue, dizziness and, in extreme cases, claustrophobia. It recommended that the operator should have frequent rests.[5]

In 1976, an American report concluded that, based on measurements and the current standards together with the present knowledge of biological effects, the VDU did not present any occupational ocular radiation hazard.[6] However, a report from a 1985 Stockholm conference attended by 1200 participants, 'Scientists look again at VDU research', accepted that VDU screen operators suffered aches, pains and sudden bouts of sleep.

Some unions, like ASTMS in the UK, have agreements with individual employers that VDU operators who become pregnant will be given alternative work. Swedish trade unions specify rest periods and other safeguards for all operators.[7] Far more important would be to insist on the redesign of the equipment. In Britain, trade-union concern has grown and many unions produce check lists for the installation and use of VDUs. These check lists were based originally on the recommendations of the International Federation of Commercial, Clerical and Technical Employees. They recommend regular eye checks at six-monthly intervals, specify frequency rates, luminance of the characters on the screen, character size, shape, form and height ratios. They also cover matters such as ambient lighting.

Far less research is devoted to the more subjective concerns of workers using VDUs. Journalists, for example, complained that the equipment gives them a feeling of isolation.[8] Managers using an electronic office system established by Citibank in New York regarded the software as 'hostile' when using advanced management work stations. When a redesign of the work station was undertaken, the philosophy was to keep existing procedures as they were and to build an electrical analogue of them. In this way, it is said, the receptivity of the users was greatly increased.[9]

More dramatic reactions by organised workers have been reported. In Norway, workers at NEBB made it quite clear to the management that they would ban a range of terminals the company intended purchasing because these could only be operated in a mode which was 'unidirectional', and hence not really responsive to the human being. Such a system, they pointed out, would be inherently undemocratic and was therefore unacceptable.

The employer purchased a different range of terminals as a result of the direct industrial and collective strength of these workers. It is quite conceivable that these workers would, in any case, have had a constitutional right to insist on such changes. An act in Norway which has been in force for seven years requires employers to provide 'sound contract conditions and meaningful occupation for the individual employee' and 'the individual employee's opportunity for self-determination'. 'Each employer shall cooperate to provide a fully satisfactory working environment for all employees at the workplace.'[10]

Architect or Bee?

TAYLOR'S SCIENTIFIC MANAGEMENT

Central to the dehumanisation of work in the intellectual field, just as in the field of manual work, is the fragmentation of work into narrow, alienated tasks, each minutely timed. To reduce the worker to a blind, unthinking appendage of the machine is the very essence of 'scientific management'. Paradoxically, Taylor's scientific management, applied to the shop floor, initially increased the intellectual activity of the staff in the offices. In his book *Shop Management*, Taylor explained that his system 'is aimed at establishing a clear cut and novel division of mental and manual work throughout the workshops. It is based upon the precise time and motion study of each worker's job in isolation, and relegates the entire mental parts of the task in hand to the managerial staff.'

Timely warnings of these dangers came from nineteenth-century writers. 'To subdivide a man is to assassinate him. The subdivision of labour is the assassination of a people.'[11]

The notion of the division of labour and the efficiency which is said to flow from it is normally associated with Adam Smith.[12] In fact, Adam Smith's specific arguments were anticipated by Henry Martyn almost a century earlier.[13] However, the basic notion of the division of labour is so intertwined with Western philosophy and scientific methodology that it is identifiable as far back as Plato when he argues for political institutions of the republic on the basis of the virtues of specialisation in the economic sphere.

The division of labour and fragmentation of skills is of course absolutely rational if you regard people as mere units of production and are concerned solely with the maximisation of the profit you extract from them. Indeed, viewed from that premise, it is not merely rational but also scientific. The scale and nature of the deskilling which accompanies this scientific management has been graphically described by Braverman.[14] This deskilling stretches right through the intellectual field. One researcher who has examined the effects of automation in Swedish banks states, 'Increased automation converted tellers, who were in effect mini-bankers, into automatons.'[15]

It might be argued in defence of these developments that at least in the 'occupational growth areas' associated with computing, those workers concerned with issuing instructions to the machines will be undertaking work of growing skill and creativity. To suggest this would be to fail completely to understand the historical tendency to deskill *all* work. Programming is itself being reduced to routines and 'the deskiller is deskilled' as structural programming breaks with the universal (if short) tradition of idiosyncratic software production.[16]

The use of this scientific management has seen the fragmentation of work occurring through the spectrum of workshop activity engulfing even the most creative and satisfying manual jobs (such as toolmaking). We are now, in addition, experiencing the same fragmentation in nonmanual jobs.

Up to the 1970s, most industrial laboratories, design offices and administrative centres were the sanctuaries of the conceptual planning and administrative aspects of work. In these areas, one spur to output was a dedication to the task in hand, an interest in it, and the satisfaction of dealing with a job from start to finish. Some observers, including the author, cautioned that the situation would soon be brought to an end as the monopolies, in their quest for increased profits, would bring their 'rational and scientific' methods into these more self-organising and comparatively easy-going fields. The objective circumstances for this were already set when in some industries 50 or 60 per cent of those employed were scientific, technical and managerial staff.

It was evident that the more science ceased to be an amateur gentleman's affair and was integrated into the productive processes, the more scientists and technologists would become part of the work force itself. It was even suggested that as high-capital equipment such as computers became available to scientists and technologists, they would be paced by the machine. Eventually, their intellectual activity would be divided into routine tasks and work study would be used to set precise times for its synchronisation with the rest of the 'rational procedure'.

Those scientists and technologists, particularly in the computer field, who look upon this view with derision, would be well advised to recall what the father of their industry, Charles Babbage, had to say on the matter. As early as the 1830s he anticipated Taylorism in the field of intellectual work. In a chapter entitled 'On the Division of Manual Labour' his message is clear: 'We may have already mentioned what may perhaps appear paradoxical to some of our readers, that the division of labour can be applied with equal success to mental as well as mechanical operations, and that it ensures in both the same economy of time.'[17]

THERBLIGS AND YALCS

In spite of these warnings and in spite of strikes by some white-collar unions against the use of the stopwatch in offices, these predictions were for the most part treated either as the scaremongering of slick trade-union leaders keen on increasing their flock, or as plain absurdity. 'That will be the day when someone tries to measure *my* intellectual activity' was a frequent reaction. Unfortunately, the day may be much closer than many would like to believe. In June 1974, there appeared in the publication *Workstudy*, 'A Classification and Terminology of Mental Work'. It suggests that much 'progress' has been made in this direction. Having identified the hierarchy of physical work – i.e. job, operation, element, therblig, it states:

> The first three of these are general concepts – i.e. they can be applied equally well to physical or mental work. The last term, the therblig, is specific to physical work. All elements of physical work consist of a small number of basic physical motions first codified by Gilbreth [Therblig is an anagram of Gilbreth] and later amended by the American Society of Mechanical Engineers and in the British Standard Glossary. The logical pattern would be complete if a similar breakdown of elements into basic mental motions – or Yalcs – were available. [Yalc is named after Clay.]

The paper describes how to classify yalcs into input, output and processing yalcs, and also how each of these can be subdivided into basic mental operations. It even draws a distinction between 'seeing', or the passive reception of visual signals, and 'looking', i.e. their active reception. Similarly it distinguishes between 'hearing', or the passive reception of audio signals, and 'listening', i.e.

their active reception. The paper implies that these techniques will be used in the more simple aspects of mental work. However it concludes by saying:

> We have tried to show that mental work is a valid and practical field for the application of workstudy; that basic mental motions exist and can be identified and classified in a meaningful way provided one does not trespass too far into the more complex mental routines and processes. A set of basic mental motions have been identified, named, described and coded as a basis for future work measurement research leading to the compilation of standard times. There are good prospects that such times could play a valuable part in workstudy projects.

It is clear, however, that these techniques *will* 'trespass too far' into the more complex mental routines and processes, just as they have in the case of highly creative manual work. Whether one regards this type of research as pseudoscientific or not, there can be little doubt about how it will be deployed. The employers of scientific, technical and administrative staff, including some forms of managerial staff, will see it as a powerful form of psychological intimidation to mould their intellectual workers to the 'mental production line'. It is perhaps a recognition of this tactical importance which prompted Howard C. Carlson, a psychologist employed by General Motors, to say: 'The computer may be to middle management what the assembly line is to the hourly paid worker.'[18]

'OBJECTIVE' SCIENTIFIC DECISION-MAKING

The computer is not only used as a Trojan horse for Taylorism in the fields of management and scientific work, even the university is no longer a sanctuary for non alienated work. Those academics engaged in the physical and pure sciences will be pleased to learn that these important issues of efficiency and optimisation will not be left to the subjective ramblings of the sociologist or the tainted ideology of the political economist. The full analytical power and neutrality of real science and the penetrating logic of mathematical method have been brought to bear. They will undoubtedly produce a completely 'objective' solution to the problem of university efficiency. For example, the notion of utilising factory models to optimise university and polytechnic productivity has been seriously proposed. We now have the rather ironic development in which some of those who, at university, helped to develop the scientific management production systems which made work so grotesque for those on the shop floor, may soon be the victims of their own repressive techniques. An article entitled 'College of Business Administration as a Production System'[19] is symptomatic of a general tendency. This article employs the terminology which describes academic features and activities in the form of a factory model. It is strongly indicative of the underlying philosophy.

Thus the recruitment of students is referred to as 'material procurement', recruiting of faculty as 'resource planning and development', faculty research and study as 'supplies procurement', instructional-methods planning as 'process planning', examinations and award of credits as 'quality control', instructor evaluation as 'resource maintenance' and graduation as 'delivery'. The professors and lecturers are of course 'operators' and presumably, as on the factory floor, only the effective operators will be tolerated. (Effective for what, and for whom, we may ask.)

The administrators' definition of effectiveness and competence makes it highly likely that many of the cherished academic freedoms of the university, whether real or imaginary, will be dented. In the not too distant future, many faculty members may well find themselves subordinated to the process in the interests of efficiency, as are workers on the shop floor. To get down to the real 'science' of it we can look at the proposals of Geoffrion, Dyer and Freiberg in 'An Interactive Approach to Multicriterion Optimisation, with an Application to the Operation of an Academic Department'.[20] They use the well known Frank-Wolfe algorithm and suggest that the multicriterion problem be reduced to the following expression:

Maximise $U [f_1(x), f_2(x), \ldots\ldots f_r(x)]$, subject to $x \in E$ where $f_1, \ldots\ldots f_r$ are r distinct criterion functions of the decision vector x, X is the constrained set of feasible decisions, and U is the decision maker's overall preference function defined on the values of criteria.

Taking a specific department as an example, they define six criteria for it. The first three are the number of course sections offered by the department at graduate, lower-division undergraduate and upper-division undergraduate levels. Criterion four is the amount of teaching-assistant time used for the support of classroom instruction by the faculty. The fifth criterion is the regular faculty effort devoted to major departmental duties measured in equivalent course sections. Finally, criterion six is the regular faculty effort devoted to additional activities such as research, student counselling and minor administrative tasks, again measured in course sections.

Terms such as 'teaching time', 'teaching loads' and 'faculty effort' are used throughout. This will mean that whoever makes a decision about criterion weight must have very precise times for the different functions, and thus the basis is clearly set for work measurement not unlike that on the shop floor. The justification will undoubtedly be that such times are necessary to be fed into the computerised model for objective assessment.

However, despite the veneer of mathematical objectivity, it is the subjective judgement of the so-called decision-maker that determines the key U function. This decision-maker will be an administrator, not the academic staff themselves, who will consequently experience a loss of control over their work environment. If, for example, a faculty member is informed via the computer that he or she is taking too long on teaching or spending too much time in research, or has been rendered superfluous as a result of an optimisation routine (a function which mathematically illiterate workers call 'the sack'), it will be worth recalling that it is the U function that predominates.

In furtherance of this efficiency, a comprehensive faculty-activity analysis was prepared and developed by the University of Washington.[21] The percentage time devoted to each faculty activity is requested. All university activities, whether regular or irregular, are refined and coded. For example, code 501 (unscheduled teaching) includes thesis-committee participation, discussion with colleagues about teaching, guest lecturing in other faculty members' courses and giving seminars within the institution. Each activity is specified very precisely as it might be in a factory situation. Under the code 'Specific Scholarly Project' are listed: departmental research, sponsored research, writing or developing research proposals, writing books and articles and many others. Under 'General Scholarly Projects' we find: reading articles and books related to the profession, attending professional meetings, research-related discussions with colleagues, and reviewing colleagues' research work.

SOME CONSEQUENCES

There are some academics who hope that in projects of this kind educational requirements will outweigh mere productivity ones, but many feel the outcome will be a shrinking of facilities, as in the City University of New York where 700 faculty members were sacked.[22] In the United States, these programmes are in fact spreading rapidly as indicated by the scale of recent grants. In the California State university and college system a 'Centre for Professional Development' was set up with a grant of $341,261 from the Fund for the Improvement of Post Secondary Education in Washington. There is no doubt what the term 'improvement' is intended to imply.

This increased productivity, however, could have consequences much more widespread and subtle than the obvious ones of increased work tempo, loss of control, job insecurity and even redundancy. The impact this will have on the creativity of those involved is likely to be significant, for central to all optimisation procedures of this kind is the notion of specific goal objectives.

A vivid example of the need to avoid such an overconstrained work environment was the design of EMI's computer-controlled brain and body X-ray scanner. In his evidence to the Select Committee on Science and Technology, Dr John Powell, EMI's managing director, pointed out that the scanner was developed using unallocated funds as a by-product of work on optical character recognition. Dr Powell stated that had its inventor 'been constrained to follow a set objective on contract research funded by an operating division, he might have just produced another optical character-recognition machine'.

The Potential and the Reality

The inventor himself, Dr Godfrey Hounsfield, who received the 1979 Nobel Prize for medicine as a result of his work on the scanner, said about his habit of going for long walks, 'It is a time when things come to one, I find. The seeds of what happened came on a ramble.' He said also, 'I still feel quite a lift when I find that the machine is doing good.'[23]

Scientific and technical advance, in spite of its liberatory potential, brings also in its wake powerful tendencies of control and authoritarian organisational forms. Indeed, it has been suggested that 'control' has been as much a stimulus to technological change as has 'productivity'.[24] Some researchers pointed out as early as the fifties and sixties that computers increase the authoritarian control which an employer has over his employees, and strengthens the hand of those who support a tougher attitude to employees.[25]

The process is succinctly described by a writer fresh from an IBM customer-training course:

> Now an operating system is a piece of software functionally designed to do most efficiently a particular job – or is it? It gradually dawned on me that some rather obnoxious cultural assumptions have been imported lock, stock and barrel into IBM software. Insidious, persuasive assumptions which appear to be a natural product of logic – but are they?
>
> The whole thing is a complete totalitarian hierarchy. The operating system runs the computer installation. The chief and most privileged element is the 'Supervisor'. Always resident in the most senior position in the main storage, it controls, through its minions, the entire operation. Subservient to the Supervisor is the bureaucratic machinery – job management routines, task management, input/output scheduling, spares management and so on. The whole thing is thought out as a rigidly controlled, centralised hierarchy, and as machines get bigger and more powerful, so the operating system grows and takes more powers.
>
> One lecturer soared into eloquence in comparing the various parts of the operating system to the directors, top management, middle management, shop foremen and ordinary pleb workers of a typical commercial company. In fact, the whole of IBM

terminology is riddled with class expressions such as master files, high and low level languages, controller, scheduler, monitor.[26]

The same writer then generalised some of the contradictions of centralised operating systems. These coincided closely with my own findings when I investigated the contradictions in the specific field of computer-aided design.

The drawbacks of the centralised operating system are many. It is a constraining and conservative force. A set of possibilities for the computer system is chosen at a point in time and a change involves regeneration of the system. It imposes conformity on programming methods and thought. Another amazingly apt quote from an IBM lecturer was 'always stick to what the system provides, otherwise you may get into trouble'. It mystifies the computer system by putting its most vital functions into a software package which is beyond the control and comprehension of the applications engineer, thus introducing even into the exclusive province of data processing the division between software experts and other programmers, and reinforcing the idea that we do not really control the tools we use, but can only do something if the operating system lets you – a phrase which I am sure many of us have used. The system which results seems absurdly top heavy and complex. The need to have everything centrally controlled seems to impose an enormous strain.

Six
POLITICAL IMPLICATIONS OF NEW TECHNOLOGY

MALE/FEMALE VALUES

Niels Bjorn-Andersen and his colleagues of the information-systems research group at the Copenhagen School of Economics have named their latest joint computerisation project with the trade unions Daphne.

The name is an acronym in Danish, but it has a much more profound significance. You may recall that in Greek mythology Daphne was a nymph, the daughter of the river Peneius. She was the embodiment of what we would nowadays refer to as the historically determined 'female' characteristics, such as intuition, subjectivity, tenacity and compassion.

She was pursued by Apollo, the embodiment of the so-called male characteristics: logic, analysis, rationality, objectivity. Indeed, one might say, the god of computerisation.

When he failed to win Daphne's favours, Apollo applied the male logic of 'might is right' and decided to take her by force.

As he was about to rape her, she called on the venerable Gaea to help her. Immediately, the earth opened, Daphne disappeared, and in her place a laurel tree sprang from the ground.

Believing that male values have raped science and technology for long enough, Bjorn-Andersen pointed out that 'it was natural for us to choose the name Daphne'.

One of the major problems with Western science and technology is that they have the historically determined male values built into them. These are the values of the white male warrior, admired for his strength and speed in eliminating the weak, conquering competitors and ruling over vast armies of men who obey his every instruction. He makes decisions which are logical, rational and will lead to victory. Within this, there is little place for the attributes of Daphne.

The introduction of a computerised system is frequently used as a smokescreen to introduce a management control weapon which discriminates against women, that of job evaluation. Pseudo-scientific reasons are given for fragmenting jobs and slotting the subdivided function into a low level of the system's hierarchy with correspondingly low wages for 'appropriate' job grades. My experience of this in industry tends to show that it is frequently used to consolidate the unequal pay and opportunities for women. This is done either by implying, or by ensuring by structural means and recruitment, that the fragmented functions are women's work. This of course can no longer be stated openly since there is the sex-discrimination legislation to watch out for, but it still happens that women are recruited for the input of predetermined data, for example, whereas the higher-status jobs are offered to men.

When we looked over some past issues of the computing magazines covering a period of six months in 1983, 82 per cent of the advertisements that had one person in a photograph with the equipment showed a woman in some kind of absurd posture which was in no way related to the use of the equipment. There is a continual projection of the view, even in the most serious of journals, that women are to be regarded as playthings, draped around the place for decoration.

Not only that, but those who read these journals often do not notice the built-in assumption unless it is pointed out to them. They are conditioned to accept the presence of women in the servicing role, and the absence of women in the organising role, as being quite normal. Even women themselves quite often see nothing untoward in this.

Technological change is starved of values like intuition, subjectivity, tenacity and compassion. It would be an enormous contribution to society if more women were to come into the technological field, not as imitation men or honorary males, but to challenge the 'male' values which have distorted it for so long. It would be a contribution to science itself which would become more caring, liberatory, socially relevant and natural.

Women are going to have to fight, not only the traditional forms of discrimination, but much more sophisticated and scientifically structured ones. There is little indication, even in 1987, that the unions catering for women workers in scientific, administrative and medical occupations which are being restructured around computer-based equipment have really understood the nature and the scale of this problem. However, what we can do is change our attitude to these 'male' and 'female' values and thereby cease to place the objective above the subjective, the rational (mathematical) above the tacit (there are things we know but cannot tell) and the digital above analogical representation.

IS SCIENCE NEUTRAL?

Marxist critics of capitalist society have tended to concentrate, at least since the turn of this century, on the contradictions of distribution. This they have done at the expense of a thoroughgoing analysis of the contradictions of production within technologically advanced society.

This imbalance can hardly be attributed to a one-sidedness on Marx's own part. Central to volume I of *Capital* is the nature of the labour process and a 'critical analysis of capitalist production'. In this, Marx demonstrates that with the accumulation of capital – the principal motivating force – the processes of production are incessantly transformed. For those who work, whether by hand or brain, this transformation shows itself as a continuous technological change within the labour process of each branch of industry, and secondly, as dramatic redistributions of labour among occupations and industries.

That the overall development of production since then should accord so closely with Marx's analysis is a remarkable tribute to his work, bearing in mind the sparsity of occupations and industries then, compared with the proliferation of these today. Whether this Marxist analysis will be equally consistent and valid when applied to the science-based industries which have emerged since the Second World War is now a matter of considerable discussion. With the integration of science into the 'productive forces' this question is one of growing significance. In some large multinational corporations 50 per cent or more of all those employed are scientific, technical or administrative 'workers'. This has begun to question, in a very practical way, the relationship between science, as at present practised, and society.

THE USE/ABUSE MODEL

Up to the mid sixties, there hardly seemed any useful purpose in raising this question. At that time, there was hardly a chink in the Bernalian analysis of twentieth-century science. In this analysis, science, although it was integral to capitalism, was ultimately in contradiction with it. Capitalism, it was felt, continuously frustrated the potential of science for human good. Therefore, the problems thrown up by the application of science and technology were viewed simply as capitalism's misuse of their potential. The contradictions between science and capitalism were viewed as the inability of capitalism to invest adequately, to plan for science, and to provide a rational framework for its widespread application in the elimination of disease, poverty and toil.

The forces of production, in particular, science and technology, were viewed as ideologically neutral, and it was considered that the development of these forces was inherently positive and progressive. It was held that the more these productive forces – technology, science, human skill, knowledge and abundant 'dead labour' (fixed capital) – developed under capitalism, the easier the transition to socialism would be. Further, science is rational, and could therefore be counterposed against irrationality and suspicion.

Science had after all, through the Galilean revolution, destroyed the earth-centred model of the universe, and, through Darwin, had made redundant earlier ideas of the creation of life and of humanity. Science, viewed thus, appeared as critical knowledge, liberating humanity from the bondage of superstition – a superstition which, elaborated into the system of religion, had acted as a key ideological prop of the outgoing social order.[1] The past few years have seen a growing questioning of this rather mechanistic interpretation of the Marxist thesis. There is now a growing realisation that science has embodied within it many of the ideological assumptions of the society which has given rise to it. This in turn has resulted in a questioning of the neutrality of science as at present practised in our society. The debate on this issue is likely to be one of major political significance. The question extends far beyond that of scientific abuses, to the deeper considerations of the nature of the scientific process itself. Science done within a particular social order reflects the norms and ideology of that social order. Science ceases to be seen as autonomous, but instead as part of an interacting system in which internalised ideological assumptions help to determine the very experimental designs and theories of scientists themselves.[2]

Failure to deal with these questions will mean that the anti-science movement of the 1970s, which had its antecedents in the anticulture movement of the 1960s, will not be developed beyond its initial and partly negative premise. In this, science is viewed as evil, totalitarian and devoid of those attributes which make it amenable to the 'human spirit'. This total rejection is now common among many young people. Indeed, in the early seventies the student population in the USA included the following words among those terms it regarded as 'bad': verification, facts, technology, statistical controls, programming, calculate, objectify, detachment.[3]

Not surprisingly, many of these students opt for the arts or social sciences where they feel (sometimes mistakenly) that more opportunity will exist for humanistic concerns.

Our Western scientific methodology is based on the natural sciences. Within this, relationships are mathematically quantifiable. There has been a tendency to suggest that if you cannot quantify something it really doesn't exist. This is not without its political significance, for if the mass of ordinary people are incapable of providing 'scientific reasons' for their judgements (which are based on actual experience of the real world), ruling elites can then silence their common sense with quantification. This has caused the brilliant French mathematician, Professor Jean-Louis Rigal, to observe, 'Quantification is the ultimate form of fascism.' Rigal's concern about quantification is even more relevant when applied outside the domain in which it evolved. Attempts to use this narrow, mathematically-based science in the much more complex and indeterminate social sciences and political activity give rise to very serious distortions, which are inevitable from the abstracted nature of the scientific method.

It is significant that those working in the scientific field are themselves beginning to raise these questions. Thus, Professor R. S. Silver says that there are risks

> in the scientific method, which may abstract common features away from concrete reality in order to achieve clarity and systematisation of thought. However, within the domain of science itself, no adverse effects arise because the concepts, ideas and principles are all interrelated in a carefully structured matrix of mutually supporting definitions and interpretations of experimental observation. The trouble starts when the same method is applied to situations where the number and complexity of factors is so great that you cannot abstract without doing some damage, and without getting an erroneous result.[4]

Those working in the field of cybernetics have also expressed their concern about this misuse of 'science'. 'There is no doubt that a very important influence nowadays is a revised reductionism within the theory of cybernetics. It reduces processes and complex objectives to black boxes and dynamic control systems. Not only in the natural sciences, but also in the social sciences.'[5]

To address these problems it will be necessary to challenge the idea of what constitutes scientific development. The role of science and technology in society will need to be recast and a social structure provided which will be capable of nurturing the coexistence of the subjective and the objective, of tacit knowledge based on contact with the physical world, and abstracted knowledge. More simply, a society and a culture which would reduce and gradually eliminate the divisions between hand and brain, and provide the stimulus, encouragement and infrastructure to permit human beings to develop in a well-rounded and heuristic fashion. This will mean challenging the fundamental assumptions of our present society and, indeed, the assumptions of societies in the so-called socialist countries. One of the important factors now moulding the social forces to give rise to such a challenge is the contradictions of science and technology experienced by an ever increasing section of the population.

CONTROL THROUGH TECHNOLOGY

The elitist right of the scientific worker or researcher to give vent to his or her creativity will now be increasingly curbed by the system as it seeks to control human behaviour in all its aspects. This is part of the general attempt of the small elite who control society to gain complete control over all those who work, whether by hand or by brain, and to use scientific management and notions of efficiency as a vehicle for doing so. It will be seen, then, that the organisation of work, and the means of designing both jobs and the machines and computers necessary to perform them, embody profound ideological assumptions. So, by regarding science and technology as neutral, we have

> failed to recognise as antihuman, and consequently to oppose the effects of values built into the apparatus, instruments and machines of their capitalist technological system. So, machines have played the part of a Trojan horse in their relation to the Labour movement. Productivity becomes more important than fraternity. Discipline outweighs freedom. The product is in fact more important than the producer, even in countries struggling for socialism.[6]

It has been suggested[7] that by ignoring these considerations the Soviet Union was laying the basis for the present situation in which it would be hard to argue that a worker there enjoys the sense of fulfilment through his or her work envisaged by the early Marxists. It may well be that in merely trying to adapt forms of science and technology developed in the capitalist societies instead of developing entirely different ones, the Soviet Union has made a profound error. The development in that country must find part of its origins in the attitude of Lenin to Taylorism, which, he said,

> like all capitalist progress is a combination of the refined brutality of bourgeois exploitation, and a number of the greatest scientific achievements in the field of analysing mechanical

motions during work, the elimination of superfluous and awkward motions, the elaboration of the correct methods of work, the introduction of the best system of accounting and control etc. The Soviet Republic must at all costs adapt all that is valuable in the achievement of Science and Technology in this field. The possibility of building socialism depends exactly on our success in combining the Soviet Power and the Soviet Organisation of Industry with the up-to-date achievements of capitalism. We must organise in Russia the study and teaching of the Taylor system, and systematically try it out and adapt it to our ends.[8]

Socialism, if it is to mean anything, must mean more freedom rather than less. If workers are constrained through Taylorism at the point of production, it is inconceivable that they will develop the self-confidence and the range of skills, abilities and talents which will make it possible for them to play a vigorous and creative part in society as a whole.

So it is that, in the technologically advanced nations, there are now beginning to emerge a range of contradictions which will necessitate a radical examination of how we use science and technology, and how knowledge should be applied in society to extend human freedom and development.

TECHNOLOGICAL CHANGE AND PROLETARIANISATION

The emergence of fixed capital as a dominant feature in the productive process means that the organic composition of capital is changed and industry becomes capital-intensive rather than labour-intensive. Human beings are increasingly replaced by machines. This in itself increases the instability of capitalism: on the one hand capitalism uses the quantity of working time as the determining element in production, yet at the same time it continuously reduces the amount of direct labour involved in the production of commodities. At an industrial level, literally millions of workers lose their jobs and millions more suffer the nagging insecurity of the threat of redundancy. An important new political element in this is the class composition of those being made redundant. Just as the use of high-capital equipment has spread out into white-collar and professional fields, so have the consequences of high-capital equipment. Scientists, technologists, professional workers and clerical workers all now experience unemployment in a manner that only manual workers did in the past. Verbal niceties are used to disguise their common plight. A large west London engineering organisation declared its scientists and technologists 'technologically displaced', its clerical and administrative workers 'surplus to requirements' and its manual workers 'redundant'. In other words they had all got the sack. In spite of different social, cultural and educational backgrounds, they all had a common interest in fighting the closure of that plant, and they did. Scientists and technologists paraded around the factory carrying banners demanding 'the right to work' in a struggle that would have been inconceivable a few years ago. Technological change was indeed proletarianising them. In consequence of the massive and synchronised scale of production which modern technology requires, redundancies can affect whole communities. During a recession in the American aircraft industry, a union banner read, 'Last out of Seattle, please put the lights out.'

Because of this change in the organic composition of capital, society is gradually being conditioned to accept the idea of a permanent pool of unemployed persons. In the United States, the 1970s saw some 5 million people permanently out of work in spite of the artificial stimulus of the Vietnam War. It is true that some of the more recent Reagan policies have resulted in job creation in limited sectors and small businesses. However, this may be more of a transitional phenomenon than one heralding the end of mass unemployment, and has been due partly to American external financial policies and at the expense of jobs in other countries. Japan and the United States have tended to export unemployment to maintain employment at home.

We have witnessed in this country the large-scale unemployment of recent years. Unemployment is considerable in Italy, and even in the West German miracle there are sections of workers, particularly over the age of fifty, who are now experiencing long terms of unemployment and there is no sign of this being reversed. (See Figure 15.) This unemployment itself creates contradictions for

Architect or Bee?

the ruling class. It does so because people have a dual role in society, that of producers and consumers. When you deny them the right to produce, you also limit their consumption power. In an attempt to achieve a balance, efforts are now being made to restructure the social services to maintain that balance between unemployment and the purchasing power of the community. In the United States, President Kennedy spoke of a 'tolerable level of unemployment'. In Britain in the 1960s, Harold Wilson, having fuelled the fires of industry with the taxpayers' money through the Industrial Reorganisation Corporation to create the 'white heat of technological change', spoke in a typical double negative of a 'not unacceptable level of unemployment' – a remarkable statement for a so-called socialist prime minister.

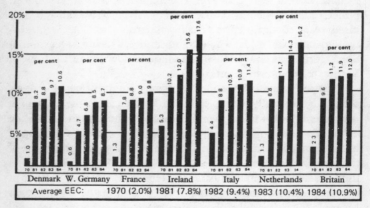

Fig. 15. Unemployment rates in the EEC.

This concept has now been extrapolated and developed by the Thatcher government to condition the population to accept that 3.5 million is a 'not unacceptable level of unemployment'. It is implied that those out of work have only themselves to blame, are scroungers or are too unimaginative, unwilling or downright lazy to avail themselves of leisure activities. Given the lack of infrastructure and resources, and the absence of job-sharing mechanisms

THE FRUITS OF EARLY OPTIMISM

with shorter working hours, those out of work experience not leisure but enforced idleness.

The net result is that there is on the one hand an increased work tempo for those in industry and on the other hand a growing dole queue with all the degradation that implies. Nor has the actual working week been reduced during this period. In spite of all the technological change since the war, the working week in Britain for those who have the jobs is now longer than it was in 1946, if we include overtime, moonlighting and travel time. Yet the relentless drive goes on to design machines and equipment which will replace workers. Those involved in such work seldom question the nature of the process in which they are engaged. Why, for example, the frantic efforts to design robots with pattern-recognition intelligence when we have three million people in the dole queue in Britain whose pattern-recognition intelligence is infinitely greater than anything yet conceived even at a theoretical level?

The policies of the labour and trade-union movement have in the past been to accept redundancies and to cut expenditure on, for example, defence without any concrete proposals whatsoever about alternative work. The argument in support of this has been that defence cuts would release capital, which could then be used in the social services. It is of course then grudgingly admitted that there would be the residual problem of further unemployment.

This reveals the extent to which those of us in the labour movement have been conditioned by the criteria of the market economy. We see the freeing of capital as an asset, and the freeing of people as a liability. In doing this we ignore our most precious asset – people, with their skill, ingenuity and creativity. In the defence and aerospace industries we have some of the most highly skilled and talented workers in Britain. Yet, like the ruling class, we have thought of capital first and people last, and ignored the great contribution which their skill and ability could make to the wellbeing of the people.

Confronted with these contradictions, the bleating and whimpering of the European trade-union bureaucracies (to be contrasted with the creative Luddism of at least some sections of the Australian movement) has failed to disguise the reality that they have no independent view of how science and technology should develop. Indeed, when they are not demanding more investment in the same forms of technology that have given rise to the problems in the first place, they are making minor, pathetic, window-dressing modifications to the proposals of the vast multinational corporations. A more constructive trade-union response is illustrated in Figure 16.

Given the gradual incorporation of the trade unions through membership of state planning bodies, industrial sector strategy teams, co-determination in West Germany, and, in Britain, quangos (although this trend has been halted and somewhat reversed under Thatcher), this form of response is perhaps not so surprising. What is, however, disconcerting is the total disarray and confusion of the Marxist Left as the political pigeons of blind unthinking technological optimism come home to roost with a vengeance.

'USING' PEOPLE

The system seeks in every way to break down workers' resistance to being sacked. One of the sophisticated devices was the Redundancy Payments Act under the Labour government. Practical experience of trade unions in Britain demonstrates that the lump sums involved broke up the solidarity at a number of plants where struggle was taking place against a closure.

A much more insidious device is to condition the workers into believing that it is their own fault that they are out of work, and that they are in fact unemployable. This technique is already widespread in the United States, where it is asserted that certain workers do not have the intelligence and the training to be employed in modern technological society. This argument is particularly used against coloured workers, Puerto Ricans and poor whites. There is perhaps here fertile ground for some of the 'objective research' of Jensen and Eysenck.

The concept of a permanent pool of unemployed persons, as a result of technological change, also brings with it the danger that

Political Implications of New Technology

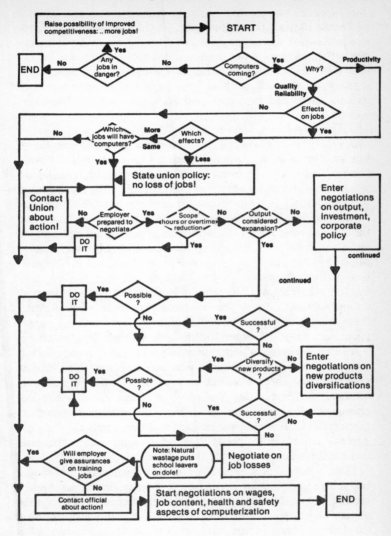

Fig. 16. A typical trade-union response to new technology.

Architect or Bee?

those unemployed would be used as a disciplining force against those still in work. It undoubtedly provides a useful pool from which the army and police force can draw. During the redundancies in Britain throughout the seventies, as the traditional industries were restructured or eliminated altogether, a considerable number of redundant workers from the northeast were recruited into the army and then deployed against workers in Northern Ireland.

Coupled with the introduction of high-capital equipment is usually a restructuring known as 'rationalisation'. The epitome of this in Britain is the General Electric Company complex with Arnold Weinstock at its head. In 1968, this organisation employed 260,000 workers and made a profit of £75 million. In consequence of quite brutal redundancies, the company's work force was reduced to 200,000 yet profits went up to £105 million. These are the kind of people who are introducing high-capital equipment, and they make their attitude to human beings absolutely clear: profits first and people last. I quote Arnold Weinstock not because he is particularly heinous (he is in fact extremely honest, direct and frank) but because he is prepared to say what others think. He said on one occasion, 'People are like elastic, the more work you give them, the more they stretch.' We know, however, that when people are stretched beyond a limit, they break. The AUEW-TASS has identified a department in a west London engineering company where the design staff were reduced from thirty-five to seventeen and there were six nervous breakdowns in eighteen months. Yet people like Weinstock are held up as a glowing example to all aspiring managers. One of his senior managers once proudly said, 'He takes people and squeezes them till the pips squeak.' I think it is a pretty sick and decaying society that will boast of this kind of behaviour.

Most industrial processes, however capital-intensive they might be, still require human beings in the total system. Since highly mechanised or automated plant frequently is capable of operating at very high speeds, employers view the comparative slowness of the human being in his interaction with the machinery as a bottleneck in the overall system. In consequence of this, pay structures and productivity deals are arranged to ensure that the workers operate at an even faster tempo.

For the employer it is like having a horse or dog. If you must have one at all then you have a young one so that it is energetic and frisky enough to do your bidding all the time. So totally does the employer seek to subordinate the worker to production that he asserts that the worker's every minute and every movement 'belong' to him, the employer. Indeed, so insatiable is the thirst of capital for surplus value, that it thinks no longer in terms of minutes of workers' time, but fractions of minutes.

The methods may vary from company to company, or from country to country, but where the profit motive reigns supreme, the degradation and subordination of the worker is the same. George Friedmann, that shrewd observer of industrial politics, has written of two different methods used by great French companies, Berliot in Lyons and Citroën in Paris:

> Why has the Berliot works the reputation, in spite of the spacious beauty of its halls, of being a jail?
>
> Because here they apply a simplified version of the Taylor method of rationalising labour, in which the time taken by a demonstrator, an 'ace' worker, serves as the criterion imposed on the mass of workers. He it is who fixes, watch in hand, the 'normal' production expected from a worker. He seems, when he is with each worker, to be adding up in an honest way the time needed for the processing of each item. In fact if the worker's movement seems to him to be not quick or precise enough, he gives a practical demonstration, and his performance determines the norm expected in return for the basic wage. Add to this supervision in the technical sphere the disciplinary supervision by uniformed warders who patrol the factory all the time and go as far as to push open the doors of the toilets to check that the men squatting there are not smoking, even in workshops where the risk of fire is nonexistent.
>
> At Citroën's the methods used are more subtle. The working teams are in rivalry with one another, the lads quarrel over travelling cranes, drills, pneumatic grinders, small tools. But the

Architect or Bee?

supervisors in white coats, whose task is to keep up the pace, are insistent, pressing, hearty. You would think that by saving time a worker was doing them a personal favour. But they are there, unremittingly on the back of the foreman, who in turn is on your back; they expect you to show an unheard of quickness in your movements, as in a speeded-up motion picture! Within this context, the desire of companies to recruit only those under the age of thirty can be seen in its dehumanised context.[9]

Although this is the position on the workshop floor, it would be naive indeed to believe that the use of high-capital equipment will be any more liberating in the fields of clerical, administrative, technical, scientific and intellectual work.

Some scientists and technologists take the smug view that this can only happen to manual workers on the shop floor. They fail to realise that the problem is now at their own doorstep. At a conference on robot technology at Nottingham University in April 1973, a programmable draughting or design system was accepted by definition as being a robot. One of the manufacturers of robotic equipment pointed out, 'Robots represent industry's logical search for an obedient workforce.' This is a very dangerous philosophy indeed. The great thing about people is that they are sometimes disobedient. Most human development, technical, cultural and political, has depended on those movements which questioned, challenged and, where necessary, disobeyed the established order.

MINIMUM MAINTENANCE FOR THE HUMAN APPENDAGE

The controllers of production view all workers, whether by hand or brain, as units of production. Only when that reality has been firmly grasped can the chasm which divides the potentialities of science and technology from the current reality be understood. The gap between what is possible and what is actual widens daily. The latent capacities of science and technology grow exponentially at the same time as the plight of many ordinary people in the West and dramatically of those in Third World countries becomes relatively worse. Technology can produce a Concorde but not enough simple heaters to save the hundreds of elderly pensioners

Political Implications of New Technology

who die in London each winter of hypothermia. Only when one realises that the system regards pensioners as discarded units of production does this make sense – capitalist sense. This is part of their social design, and from a ruling-class viewpoint it is quite 'scientific' and abides closely by the principles observed in machine design.

I know, as a designer, that when you design a unit of production you ensure that you design it to operate in the minimum environment necessary for it to do its job. You seek to ensure that it does not require any special temperature-controlled room unless it is absolutely essential. In designing the lubrication system you do not specify any exotic oils as lubricants unless it is necessary. You ensure that its control system is provided with the minimum brain necessary for it to do its job. You don't, for example have a complex CAD three-dimensional system if you can get away with a simple two-dimensional plotter. Finally, you provide it with the minimum amount of maintenance. In other words, you design for it the maximum life span in which it will operate before it fails. Those who control our society see human beings in the same way. The minimum environment for workers means that you provide them with the absolutely lowest level of housing which will keep them in a healthy enough state to do their job. The equivalent of fuel and lubrication for the machine is the food provided for a worker. This is also kept at a minimum for those who work – and is completely inadequate for those who cannot work.

In the early 1970s Oxford dietitians were telling pensioners precisely how much margarine and which scraps of meat to purchase so that they could survive on £2 worth of food per week. Despite the stir this caused at the time, the amount which working-class pensioners have available today for food is relatively unchanged in terms of actual purchasing power.

The minimum brain is provided for the worker by an educational system which imparts enough knowledge to be of use to the industrial system, and which trains him or her to do the job, but does not educate the worker to think about his or her own predicament or that of society as a whole.

The minimum maintenance is provided through the National

Health Service, which concentrates on curative rather than preventive medicine. The harsh reality is that when workers have finished their working life they are thrown on the scrap heap like obsolete machines.

If all this sounds like an extreme position, it is worth recalling the statement of the doctor at Willesden Hospital who said there was no need to resuscitate National Health patients over the age of sixty-five. (The doctor himself was sixty-eight.) When a barrage of protest was raised, the statement was hurriedly withdrawn as a mistake. The real mistake he made was to reveal in naked print one of the underlying assumptions of our class-divided society. Science and technology cannot be humanely applied in an inherently inhuman society, and the contradictions for scientific workers in the application of their abilities will grow and, if properly articulated, will lead to a radicalisation of the scientific community.

NEED FOR PUBLIC INVOLVEMENT

Any meaningful analysis of scientific abuse must probe the very nature of the scientific process itself, and the objective role of science within the ideological framework of a given society. As such, it ceases to be merely a 'problem of science' and takes on a political dimension. It extends beyond the important but limited, introverted soul-searching of the scientific community, and recognises the need for wider public involvement. Many 'progressive' scientists now realise that this is so, but still see their role as the interpreters of the mystical world of science for a largely ignorant mass which, when enlightened, will then support the scientists in their intention 'not to use my scientific knowledge or status to promote practices which I consider dangerous' (as correctly advocated by some members of the British Society for Social Responsibility in Science).

Those who in addition to being 'progressive' have political acumen know that a Lysistrata movement, even if it could be organised, is unlikely to terrify international capital into applying science in a socially responsible manner. Socially responsible science is only conceivable in a politically responsible society. That must mean changing the one in which we now live.

Political Implications of New Technology

One of the prerequisites for such political change is the rejection of the present basis of our society by a substantial number of its members, and a conscious political force to articulate that contradiction as part of a critique of society as a whole. The inevitable misuse of science, and its consequent impact on the lives of an ever growing mass of people, provides the fertile ground for such a political development. It should constitute an important weapon in the political software of any conscious radical.

Even Marxist scientists seem to reflect the internal political incestuousness of the scientific community, and demonstrate in practice a reluctance to raise the issues in the mass movement. Thus the debate has tended to be confined to the rarefied atmosphere of the campus, the elitism of the learned body or the relative monastic quiet of the laboratory.

Clearly those who control the vast multinational corporations, who have never harboured any illusions about the ideological neutrality of science, will not be overconcerned by this responsible disquiet. The Geneens of ITT and the Weinstocks of GEC do not tremble at the pronouncements of Nobel laureates. It is true, of course, that the pronouncements of the ecologists have reverberated through the quality press and caused some concern – not all of it healthy – in liberal circles. But ordinary people – those who have it within their power to transform society, those for whom such a transformation is an objective necessity – have not yet been really involved. Yet their day-to-day experience at the point of production brutally demonstrates that a society which strives for profit maximisation is incapable of providing a rational social framework for technology (which they see as applied science).

Socially irresponsible science not only pollutes our rivers, air and soil, produces CS gas for Northern Ireland, defoliants for Vietnam and stroboscopic torture devices for police states. It also degrades, both mentally and physically, those at the point of production, as the objectivisation of their labour reduces them to mere machine appendages. The financial anaesthetic of the 'high-wage (a lie in any case), high-productivity, low-cost economy' has demonstrably failed to numb workers' minds to the human costs of the fragmented, dehumanised tasks of the production line. Although the organisation of work seeks to reduce them to zombies, they develop coping mechanisms, sometimes through compensation outside work in the form of hobbies, frequently through trade-union activity and making plans for the day when they will 'escape' from the production line altogether.

There are growing manifestations in the productive superstructure of the irreconcilable contradictions at the economic base. The sabotage of products on the robot-assisted line at General Motors' Lordstown plant in the US, the 8 per cent absentee rate at Fiat in Italy, the 'quality' strike at Chrysler in Britain and the protected workshops in Sweden reveal but the tip of a great international iceberg of seething industrial discontent. That discontent, if properly handled, can be elevated from its essentially defensive, negative stance into a positive political challenge to the system as a whole.

The objective circumstances for such a challenge are developing rapidly as the crushing reality is hammered home by the concrete experience of more and more workers in high-capital, technologically based, automated or computerised plants. In consequence, there is a gradual realisation by both manual and staff workers that the more elaborate and scientific the equipment they design and build, the more they themselves become subordinated to it, that is, to the objects of their own labour. This process can only be understood when seen in the historical and economic context of technological change as a whole.

FUNDAMENTAL DIFFERENCE

The use of fixed capital, that is machinery and, latterly, computers, in the productive process marked a fundamental change in the mode of production. It cannot be viewed merely as an increase in the rate at which tools are used to act on raw material. The hand tool was entirely animated by the workers, and the rate at which the commodity was produced – and the quality of it – depended (apart from the raw materials, market forces and supervision) on the strength, tenacity, dexterity and ingenuity of the worker. With fixed capital, that is, the machine, it is quite the contrary in that the method of work is geared to profit and the convenience of the machine. The scientific knowledge which predetermines the speeds and feeds of the machine, and the mathematics used in compiling the numerical control program, do not exist in the consciousness of the operator; they are external to him and act upon him through the machine as an alien force. Thus science, as it manifests itself to the workers through fixed capital, although it is merely the accumulation of the knowledge and skill now appropriated, confronts them as an alien and hostile force, and further subordinates them to the machine. The nature of their activity, the movements of their limbs, the rate and sequence of those movements – all these are determined in quite minute detail by the 'scientific' requirements of fixed capital. Thus objectivised labour in the form of fixed capital emerges in the productive process as a dominating force opposed to living labour. We shall see subsequently, when we examine concrete situations at the point of production, that fixed capital represents not only the appropriation of *living* labour but in its sophisticated forms (computer hardware and software) appropriates the scientific and intellectual output of white-collar workers whose own intellects oppose them also as an alien force.

The more, therefore, that workers put into the object of their labour, the less there remains of themselves. The welder at General Motors who takes a robotic welding device and guides its probes through the welding procedures of a car body is building skill into the machine and deskilling himself. The accumulation of years of welding experience is absorbed by the robot's self-programming systems and will never be forgotten. Similarly, a mathematician working as a stressman in an aircraft company may design a software package for the stress analysis of airframe structures and suffer the same consequences in his job. In each case they have given part of themselves to the machine and in doing so have conferred 'life', in systems terms, on the object of their labour, but now this life no longer belongs to them but to the owner of the object.

Since the product of their labour does not belong to the workers, but to the owner of the means of production in whose service the work is done, and is used in his interests, it necessarily follows that the object of the workers' labour confronts them as an alien and hostile force. Thus this 'loss of self' of the worker is but a manifestation of the fundamental contradictions at the economic base of our society. It is a reflection of the antagonistic contradiction between the interest of capital and labour, between the exploiter and the exploited. Fixed capital, therefore, at this historical stage, is the embodiment of a contradiction, namely that the means which could make possible the liberation of the workers from routine, soul-destroying, backbreaking tasks is rather the means of their own enslavement.

IS 'POLITICAL' CHANGE ENOUGH?

It will therefore be necessary to change the nature and the ownership of the means of production, although this of itself will by no means be adequate. In addition there is the question as to whether there is a contradiction (non antagonistic) between science and technology in their present form and the very essence of humanity. It is quite conceivable that our scientific methodology, in particular our design methodology, has been distorted by the social forces that have given rise to it. The question is therefore whether the problems of scientific development and technological change, which are *primarily* due to the nature of our class-divided society, can be solved solely by changing the economic base of that society.

The question is not one of mere theoretical and academic interest. It must be a burning issue in the minds of those attempting to build a people's democracy. It must be of political concern to them to establish if Western technology can be simply applied to a socialist society. Technology, at this historical stage, is the embodiment of two opposites: the possibility of freeing workers and the actuality of ensnaring them. The possibility can only become actuality when the workers own the objects of their labour. Because the nature of this contradiction has not been understood, there have been the traditional polarised views, 'technology is good' and 'technology is bad'. These polarised views are of long standing and not merely products of space-age technology. As far back as 1642, when Pascal introduced his mechanical calculating device, he expected it to free people to engage in creative work. Only forty-six years earlier, in 1596, an opposite view was dramatically demonstrated when the city council of Danzig hired an assassin to strangle the inventor of a labour-saving ribbon-loom. This reaction has been repeated time and again in various guises during the ensuing 500 years to resolve a contradiction at an industrial level when only a radical political one would suffice. That contradiction manifests itself in industrial forms even to this day.

THE DEDICATED APPENDAGE

It has been common for some time to talk about 'dedicated machines'. It is now a fact that when defining a job function, employers define a dedicated appendage to the machine, the operator.

Even our educational system is being distorted to produce these 'dedicated men for dedicated machines'. People are no longer being educated to think, they are being trained to do a narrow, specific job. Much of the unrest amongst students is caused by recognition that they are being trained as industrial fodder for the large monopolies in order to fit them into narrow fragmented functions where they will be unable to see in an overall panoramic fashion the work on which they are engaged.

In order to ensure that the right kind of 'dedicated product' is turned out of the university, we find the monopolies attempting to determine the nature of university curricula and research programmes. Warwick University was a classic example. In particular, at research level, the monopolies increasingly attempt to determine the nature of research through grants which they provide to universities or research projects undertaken in their own laboratories. Many research scientists still harbour illusions that they are in practice 'independent, dedicated searchers after truth'.

The 'truth' for them has to coincide with the interests of the monopolies if they are to retain their jobs. William H. Whyte Jr pointed out in 1960 that in the United States, out of 600,000 persons then engaged in scientific research, not more than 5000 were allowed to choose their research subject and less than 4 per cent of the total expenditure was devoted to 'creative research' which does not offer immediate prospects of profits. He recognises the long-term consequences of this and concludes, 'If corporations continue to mould scientists the way they are now doing, it is entirely possible that in the long run this huge apparatus may actually slow down the rate of basic discovery it feeds on.'[10]

PROBLEMS FOR THE EMPLOYER

I have up to now concentrated on the contradictions as they affect the worker by hand or brain. There are of course problems for the employer and an understanding of some of these is of considerable tactical importance.

One of the contradictions for the employer is that the more capital he accumulates in any one place, the more vulnerable it becomes. The more closely he synchronises his industry and production by using computers, the greater becomes the strike power of those employed in it. Mao Tse Tung once said, in his military writings, that the more capitalised an army becomes, the more vulnerable it becomes. This was demonstrated in Vietnam, where a National Liberation Front cadre with a £1.50 shell could destroy an American aircraft with an airborne computer costing something like £2.5 million.

A Palestine guerrilla with a revolver costing perhaps £20 can hijack a plane costing several million dollars and destroy it at some airfield. High-capital equipment, although it appears all-powerful and invincible, always has a point of vulnerability and possibilities for sabotage and guerrilla warfare are considerable. A quite small force can destroy or immobilise plant equipment or weapons costing literally millions. The capitalisation of industry also produces an analogous situation. In the past, when a clerical worker went on strike it had precious little effect. Now, if the wages of a factory are carried out by a computer, a strike by clerical workers can disrupt the whole plant. It is also true on the factory floor that in the highly synchronised motorcar industry a strike of twelve workers in the foundry can stop large sections of the entire industry.

The same is happening in the design area. As high-capital equipment, through computer-aided design, is being made available to design staffs, it proletarianises them, but it also increases their strike power. In the past when a draughtsman went on strike he simply put down his 6H pencil and his rubber, and there was unfortunately a considerable length of time before an effect was felt on production, even when the manual workers were blacking his drawings. With the new kind of equipment described, where control data is being prepared or where high-capital equipment is used for interactive work, the effects of a strike will in many instances be immediate, and production will be affected in a very short time.

PARITY WITH THE MACHINE

While the introduction of fixed capital enables the employer to displace some workers and subordinate others to the machine, it also embodies within it an opposite in that it provides the worker with a powerful industrial weapon to use against the employer who introduced it. This will apply equally to hosts of other jobs and occupations in banking, insurance, power generation, civil transport, as well as those more closely connected to industry and production.

This is even the case when industrial action short of strike action is taken. As I have pointed out, the activity of the worker is transformed to suit the requirement of fixed capital. The more complete the transformation, the greater is the disruptive effect of the slightest deviation by the worker from his predetermined work sequence. Industrial militants with an imaginative and creative view of industrial harassment have been able to exploit this contradiction by developing techniques like 'working to rule', 'working without enthusiasm' and 'days of noncooperation'. These techniques can reduce the output of both manual and staff workers by up to 70 per cent without placing on the workers involved the economic hardship of a full strike.

Since much of the sophisticated equipment I have described earlier is very sensitive and delicate in a scientific sense, it has to be handled with great care and is accommodated in purpose-built structures in conditions of clinical cleanliness. In many industries the care the employer will lavish on 'his' fixed capital is in glaring contrast with the comparatively primitive conditions of 'his' living capital. The campaign for parity with equipment, which perhaps started facetiously in 1964 with a placard at Berkeley which parodied the IBM punchcard ('I am a human being: Please do not fold, spindle or mutilate'), has now assumed significant industrial dimensions. In June 1973, designers and draughtsmen, members of AUEW-TASS employed by a large Birmingham engineering firm, officially claimed 'parity of environment with the CAD equipment' in the following terms:

> This claim is made in furtherance of a long-standing complaint concerning the heating and ventilation in the Design and Drawing Office Area going back to April 1972. Indeed to our certain knowledge these working conditions have been unsatisfactory as far back as 1958. We believe that if electromechanical equipment can be considered to the point of giving it an air-conditioned environment for its efficient working, the human beings who may be interfaced with this equipment should receive the same consideration.

It is an interesting reflection on the values of advanced technological society that it subsequently took three industrial stoppages to achieve for the designers conditions approaching those of the CAD equipment. The exercise also helped to dispel some illusions about highly qualified design staff not needing trade unions.

Scientists must now begin to learn the lessons of such experiences, and to understand that their destiny is bound up with all those 'moulded' by the system. They must attempt to understand that the products of their ingenuity and scientific ability will become the objects of their own oppression and that of the mass of the people until they are courageous enough to be involved in political struggle with them. It is the historical task of the working class to effect a transformation of society, but in that process scientists and technologists can be powerful and vital allies for the working class as a whole. This means that scientists will have to involve themselves in the political movement.

When the enslaving subordination of the individual to the division of labour, and with it the antithesis between mental and physical work has vanished, when labour is no longer merely a means of life but has become life's principal need, when the productive forces have also increased with the all-round development of the individual, and all the springs of cooperative wealth flow more abundantly. Only then will it be possible completely to transcend the narrow outlook of the bourgeois right, and only then will society be able to inscribe on its banners – 'From each according to his ability, to each according to his needs'. Then, and then only will scientists be able to truly give of their ability to meet the needs of the community as a whole rather than maximise profits for the few.[11]

Seven
DRAWING UP THE CORPORATE PLAN AT LUCAS AEROSPACE

At no stage in human history has the potential for solving our economic problems been so great. Human ingenuity, expressed through appropriate science and technology, could do much to free our world from squalor and disease and fulfil our basic needs of food, warmth and shelter. Yet at the same time there is a growing disquiet, even alarm, among wide sectors of the community about the future of 'industrial society'.

THE CONTRADICTIONS

There are many contradictions which highlight the problems of our supposedly technologically advanced society. Four of these contradictions in particular influenced the events at Lucas Aerospace.

Firstly, there is the appalling gap which now exists between what technology could provide for society, and what it actually does provide. We have levels of technological sophistication such that we can guide a missile system to another continent with an accuracy of a few metres, yet the blind and disabled stagger around our cities in very much the same way as they did in medieval times. We have vast nuclear power industries, huge conventional power-generating systems, complex distribution networks and piped natural gas, yet pensioners die of hypothermia because they cannot get a simple effective heater. In the winter of 1984, some 1000 died of the cold in the London area alone. We have senior automotive engineers who sit in front of computerised visual display units 'working interactively to optimise the configuration' of car bodies to make them aerodynamically stable at 120 miles an hour when the average speed of traffic through New York is 6.2 miles an hour. It was in fact 11 mph at the turn of the century when the vehicles were horse-drawn. In London at certain times of the day it is about 8.5

Drawing up the Corporate Plan at Lucas Aerospace

miles an hour. We have such sophisticated communications systems that we can send messages round the world in fractions of a second, yet it now takes longer to send a letter from Washington to New York than it did in the days of the stage coach.

We find on the one hand the linear drive forward of complex esoteric technology in the interests of the multinational corporations, and, on the other hand, the growing deprivation of communities and the mass of people as a whole.

The second contradiction is the tragic waste our society makes of its most precious asset: the skill, ingenuity, energy, creativity and enthusiasm of its ordinary people. We now have in Britain, even according to the doctored official figures, some 3.3 million people out of work. The real, if unofficial, figure must be close to 4.5 million if we take into account those who haven't registered, those who would work part time, and the thousands of women who would welcome the opportunity of a job if it could be provided on a basis of flexible time.

There are thousands of engineers suffering the degradation of unemployment when we urgently need cheap, effective and safe transport systems for our cities. There are thousands of electricians robbed by society of the right to work when we urgently need economic urban heating systems. We have, I believe, 180,000 building workers out of a job when by the government's own statistics it is admitted that about 7 million people live in slums in this country. In the London area about 20 per cent of the schools lack an adequate indoor toilet and the people who could be making these are rotting away in the dole queue.

The third contradiction is the myth that computerisation, automation and the use of robotic devices will automatically free human beings from soul-destroying, backbreaking tasks and leave them free to engage in more creative work. The experience of my trade-union colleagues and that of millions of workers in the industrial nations is that in most instances the reverse is the case.

At an individual level, the totality that is a human being is ruthlessly torn apart and its component parts set one against the other. The individual as producer is required to perform grotesque alienated tasks to make throwaway products to exploit the individual as consumer. We are at a stage where our incorporate science and technology, with their concepts of efficiency and optimisation, converge with the requirements and value systems of the vast multinational corporations.

Fourthly, there is the growing hostility of society at large to science and technology as at present practised. If you go to gatherings where there are artists, journalists and writers and you admit to being a technologist, they treat you as scum. They really seem to believe that you specified that rust should be sprayed on car bodies before the paint is applied, that all commodities should be enclosed in containers that can't be recycled, and that every large-scale plant you design is produced specifically to pollute the air and the rivers. There seems to be no understanding of the manner in which scientists and technologists are used as mere messenger boys of the multinational corporations whose sole concern is the maximisation of profits. It is therefore not surprising that some of our most able and sensitive sixth-formers will not now study science and technology because they perceive it to be such a dehumanised activity in our society.

Propelled by the frantic linear drive forward, of this form of science and technology, we witness the exponential change in the organic composition of capital and the resultant growth of massive structural unemployment. So stark is the situation becoming that predictions of 20 million jobless in the EEC countries by 1990 no longer seem absurd.

LUCAS WORKERS RESPOND

All these four contradictions impacted themselves upon the work force in Lucas Aerospace during the 1970s. We were working on equipment for Concorde, we had experienced structural unemployment and we knew, day by day, of the growing hostility of the public to science and technology.

Lucas Aerospace was formed in the late 1960s when parts of Lucas Industries took over sections of GEC, AEI and a number of other small companies. It was clear that the company would engage in a rationalisation programme along the lines already established by Arnold Weinstock in GEC. This, it will be recalled, was the

Drawing up the Corporate Plan at Lucas Aerospace

time of Harold Wilson's 'white heat of technological change'. The taxpayer's money was being used through the Industrial Reorganisation Corporation to facilitate this rationalisation programme. No account at all was taken of the social cost and Arnold Weinstock subsequently sacked 60,000 industrial workers with a wide range of skills.

We in Lucas Aerospace were fortunate in that this happened about one year before the company embarked on its rationalisation programme. We were therefore able to build up a Combine Committee which would prevent the company setting one site against the other in the manner Weinstock had done. This body, the Combine Committee (which is still active in 1987), is unique in the British trade-union movement in that it links together the highest-level technologists and the semiskilled workers on the shop floor. There is therefore a creative cross-fertilisation between the analytical power of the scientist and, what is perhaps more important, the direct class sense and understanding of those on the shop floor.

As structural unemployment began to affect us, we looked around at the way other groups of workers were attempting to resist it. In Lucas we had already been engaged in sit-ins, in preventing the transfer of work from one site to another and in a host of other industrial tactics which had been developed over the past five years, but we realised that the morale of a work force very quickly declines if the workers can see that society, for whatever reason, does not want the products that they make. We therefore evolved the idea of a campaign for the right to work on socially useful products.

It seemed absurd to us that we had all this skill and knowledge and facilities at the same time as society urgently needed equipment and services which we could provide, and yet the market economy seemed incapable of linking the two. What happened next provides an important lesson for those who wish to analyse how society can be changed.

AN IMPORTANT LESSON

We prepared a letter which described in great detail the nature of the work force, its age, its skills, its qualifications and the machine tools, equipment and laboratories that were available to us, together with the types of scientific staff and the design capabilities which they possessed. The letter went to 180 leading authorities, institutions, universities, trade unions and other organisations, all of which had in the past, one way or another, suggested that there was a need for the humanisation of technology and the use of technology in a socially responsible way. What happened was a revelation to us. All these people who had made great speeches up and down the country, and in some cases written books about these matters, were smitten into silence by the specificity of our request. We had asked them, quite simply, 'What could a work force with these facilities be making that would be in the interests of the community at large?' and they were silent – with the exception of four individuals, Dr David Elliott at the Open University, Professor Meredith Thring at Queen Mary College, Richard Fletcher and Clive Latimer, both at the North East London Polytechnic.

We then did what we should have done in the first place. We asked our own members what they thought they should be making.

I have never doubted the ability of ordinary people to cope with these problems, but not doubting is one thing, having concrete evidence is something quite different. That concrete evidence began to pour in to us within three or four weeks. In a short time we had 150 ideas of products which we could make and build with the existing machine tools and skills we had in Lucas Aerospace. We elicited this information through our shop stewards' committees via a questionnaire.

This questionnaire was very different from those which the soap-powder companies produce, where the respondent is treated as some kind of passive cretin. In our case, the questionnaire was dialectically designed. By that I mean that in filling it in, the respondents were caused to think about their skill and ability, the environment in which they worked and the facilities available to them. We also composed it so that they would think of themselves in their dual role in society, that is, both as consumers and as producers. We were, therefore, deliberately transcending the absurd division which our society imposes upon us, which seems to suggest that there are two nations, one that works in factories and offices, and an entirely different nation that lives in houses and communities. We pointed out that what we do during the day at work should also be meaningful in relation to the communities in which we live. We also designed the questionnaire to cause the respondents to think of products for their use value and not merely for their exchange value.

When we had collected all these proposals, we refined them into six major product ranges. These are now embodied in six volumes, each of approximately 200 pages. They contain specific technical details, economic calculations and even engineering drawings. We sought a mix of products which included, on the one hand, those which could be designed and built in the very short term, and, on the other, those which would require long-term development; those which could be used in metropolitan Britain mixed with those which would be suitable for use in the Third World – products that could be sold in a mutually nonexploitative fashion. Finally, we sought a mix of products which would be profitable by the present criteria of the market economy and ones which would not necessarily be profitable but would be highly socially useful.

THE PRODUCTS AND IDEAS

Before we even started the Corporate Plan, our members at the Wolverhampton plant visited a centre for children with spina bifida and were horrified to see that the only way the children could propel themselves about was by crawling on the floor, so they designed a vehicle which subsequently became known as hobcart. It was highly successful, and the Spina Bifida Association of Australia wanted to order 2000. Lucas would not agree to manufacture it because, they said, it was incompatible with their product range.

At that time the Corporate Plan was not developed and so we were unable to press for the manufacture of hobcart. However, the design and development of this product were significant in another sense. Mike Parry Evans, its designer, said that it was one of the most enriching experiences of his life when he delivered the hobcart to a child and saw the pleasure on the child's face. It meant more to him, he said, than all the design activity he had been involved in up till then. For the first time in his career he saw the person who was going to benefit from the product he had designed, and he was intimately in contact with a social human problem. He needed to make a clay mould of the child's back so that the seat would support it properly. He was working in a multi-disciplinary team together with a medical doctor, a physiotherapist and a health visitor. This illustrates very graphically that aerospace technologists are not only interested in complex esoteric technical problems. It can be far more enriching for them if they are allowed to relate their technology to really human and social problems.

A LIFE-SUPPORT SYSTEM

Some of our members at another plant realised that about 30 per cent of the people who die of heart attacks die before they reach the intensive-care unit in the hospital. They designed a light, simple, portable life-support system which can be taken in an ambulance or at the side of a stretcher to keep the patient 'ticking over' until he or she can be linked to the main life-support system in the hospital.

They also learned that many patients died during critical operations because of the problems of maintaining the blood at a constant optimum temperature and flow. This, it seemed to them, was a simple technical problem once they were able to get behind the feudal mysticism of the medical profession. They designed for this a simple heat exchanger and pumping system and they built it in prototype. I understand that when the assistant chief designer at one of our plants had to have a critical operation, they were able to convince the local hospital to use it, and it was highly successful.

ENERGY-CONSERVING PRODUCTS

In the field of alternative energy sources we came up with a very imaginative range of proposals. It seemed to us absurd that it takes more energy to keep New York cool during the summer than it does to heat it during the winter. Systems which could store this energy when it was not required and use it at a time when it was required would make a lot of sense.

One of the proposals for storing energy was to produce gaseous hydrogen fuel cells. These would require considerable funding from the government, but would produce a means of conserving energy which would be ecologically desirable and socially responsible.

There are further designs for a range of solar-collecting equipment which could be used in low-energy houses. We worked on this in conjunction with Clive Latimer and his colleagues at the North East London Polytechnic, and components for a low-energy house were produced. This house was designed so that it could be built by its owner. Some of the students working on the communications design degree course at that polytechnic have written an assembly-instruction manual based on directions given by the skilled people who designed the low-energy housing. This manual would allow people, working side by side with skilled building workers, to go through a learning process and at the same time produce very ecologically desirable forms of housing. If this concept were linked to imaginative government community funding, it would be possible, in areas of high unemployment where there are acute housing problems, to provide funds for employing those in that area to build houses for themselves.

In order to demonstrate the potential of this in practice, Clive Latimer has constructed a low-energy house in Suffolk and lives in it. The technical press were invited to view it in 1984 and were extremely enthusiastic about its potential. The London Innovation Network, one of the Greater London Enterprise Board's technology networks, is now developing the project further and a scale model was demonstrated at the Energy Exhibition in London in the autumn of 1985.

We made a number of contacts with county councils as we were keen to see these products used in communities by ordinary people. We were unhappy about the present tendency in alternative technology for products to be provided which are little more than playthings for the middle class in their architect-built houses, so we made links via the Open University with the Milton Keynes corporation and in conjunction with the OU we designed and built some prototype heat pumps for installation in the corporation's houses. These pumps use natural gas and have a coefficient of performance (COP) of 2.8 when it is 0°C outside the building.

Obviously, heat pumps have been around for many years, but they are usually electrically driven. Given the energy losses in the transfer from fossil fuel to electrical power and the transmission losses in the lines, only a little over 30 per cent of the original fossil-fuel value is ultimately available in the house or building as electrical energy to drive the pump. The real advantage of the natural-gas heat pump over electrically driven ones is that you start with some 70 per cent of the original fossil-fuel value and still get a COP of 2.8.

A NEW HYBRID POWER PACK

The problem of finding an ecologically desirable power unit for cars is one which needs to be solved as a matter of urgency.

Lucas Electrical, which is a separate company from Lucas Aerospace, proposed a solution based on a battery-driven car. However, with a vehicle of this kind it is necessary to recharge it approximately every forty miles of stop-start driving and every 100 miles on flat terrain. Furthermore, it is necessary to carry a significant weight of batteries. At a chassis weight of around 1300 kilos an additional 1000 kilos of batteries would have to be carried. Because the batteries need charging at regular intervals, vehicles of this kind are unsuitable for random journeys, and a large number of roadside charging facilities would be needed.

One possibility would be to provide these in existing garages, but having a large number of vehicles waiting to be charged overnight would create considerable difficulties. Interchangeable batteries could be made available, but it would clearly be a significant task regularly to change 1000 kilos of batteries (about a ton). Moreover, storage space for batteries would in the London area cost around £6 to £10 per square foot per annum. Drivers would have to pay for the additional reserve batteries *and* for the space to store them.

The aerospace workers' approach was quite different. They pointed out that the average vehicle has an engine twice and maybe three times larger than necessary, simply to give it take-off torque. Once the vehicle is moving along, a much smaller engine could satisfactorily power it. They also pointed out that the performance characteristics of an electric motor are the opposite of those of a petrol engine. That is to say, the electric motor has a high starting torque whereas a petrol engine has a better torque at high revs. By linking these two together, a new unity can be formed. A small component combustion engine, running constantly at its optimum revs and at its optimum temperature, drives a generator which in turn charges a very small stack of batteries. These act merely as a temporary energy store and supply power to an electric motor which drives the transmission system, or, in a revised version, will drive hub motors directly on the wheels.

A number of variations on this theme were proposed. One is for intercity driving. Once the vehicle has gained speed, the combustion engine could drive the wheels directly through the mechanical transmission system, whereas when the vehicle enters the suburban area, with the consequent stop-start driving, it could run in the hybrid mode.

The Lucas workers envisaged that in coming years the internal combustion engine would be banned from city centres. With the hybrid power pack it would be possible to drive to the perimeter of the prohibited zone and then, within the zone, drive slowly, solely on stored energy. The system would be recharged when operating in the hybrid mode elsewhere.

In general use, however, the internal combustion engine would be running continuously at its constant optimum rate. All the energy that is wasted as one starts from cold, accelerates and decelerates, changes gear or idles at traffic lights, would go into the system as useful energy. This, it is suggested, would improve fuel consumption by about 50 per cent. Since the engine is running constantly and at its optimum temperature, it follows that combustion of the gases will be much more complete, thereby reducing the emission of toxic fumes by about 80 per cent, since the unburned gases are not going out into the atmosphere, and it would improve specific fuel consumption by 50 per cent. The initial calculations on this have subsequently been supported by work done in Germany.

Architect or Bee?

The engine would run at constant revs so the resonance frequencies of the various components in the system could be different from that of the engine and noise levels reduced. With a background noise of 65 decibels, the power pack would be inaudible ten metres away. A prototype unit of this kind was built and tested under the direction of Professor Thring at Queen Mary College, London. Similar hybrids are now being developed in Germany and Japan.

No individual component of the system is in itself revolutionary. What is new is the creative manner in which the various elements have been put together. The only reason why such a power pack had not been designed and developed before, it seems to us, is that they would have to last for fifteen years or so, to accord with our views of long-life products to conserve energy and materials and to justify the installation cost, and maintenance services would have to be developed to repair and maintain them. This is completely contrary to the whole ethos of automotive design, which has as its basis the notion of a nonrepairable throwaway product with all the terrible waste of energy and materials which that implies.

While the Lucas workers are proposing this kind of power pack, their colleagues at another large car manufacturer's are having to design and develop an engine which would be thrown away after 20,000 miles or two years, whichever comes first. The idea is that the engine would simply be bolted on the input side of the gearbox so that it could simply be unbolted and replaced with another at the end of its life cycle. The owner would even be denied the pleasure of putting water and oil in. It is a criminally irresponsible type of technology, yet the whole political and economic infrastructure of society is based on the assumptions of this technology, namely that the rate of obsolescence of products will increase, and that the rates of production and consumption will grow. I am convinced that Western society cannot carry on in this wasteful and arrogant way much longer.

ALL-PURPOSE POWER GENERATION

Drawing on our aerodynamics know-how, we proposed a range of wind generators. In some instances these would have a unique rotor control in which the liquid used as the medium for transmitting the heat is also used to effect the braking, and is heated in the process.

We proposed a range of products which would be useful in Third World countries. We feel, incidentally, that we should be very humble about suggesting that our kind of technology would be appropriate in these countries. Probably one of the most important things the Third World countries could learn from us, looking at the incredible mess we have made of technology in our society, is what not to do. It is also very arrogant to believe that the only form of technology is that which we have in the West. I can see no reason why there should not be technologies compatible with the cultural and social structures of these other countries.

At the moment, our trade with these countries is essentially neocolonialist. We seek to introduce forms of technology which will make them dependent on us. When the gin-and-tonic brigade go out to sell a power pack, for example, they always seek to sell a dedicated power pack for each application; that is, one power pack for generating electricity, another power pack for pumping water and so on.

The Lucas workers' approach is quite different. They designed a universal power pack which is quite capable of providing a wide range of services. It has a basic prime mover which could run on many different fuels, including naturally available materials and methane gas.

By using a specially designed, variable speed gearbox it is possible greatly to vary the output speed. The unit is capable of providing the speed and power necessary to drive a generator which could supply electricity at night. When running at one of the lower speeds, it could drive a compressor to provide compressed air for pneumatic tools. It could drive a hydraulic pump to provide power for lifting equipment, and at very low speeds it could drive a water pump and be used for irrigation. The unit could thus be used in a number of ways for almost twenty-four hours a day.

In considering the design, the various bearing surfaces have been made much larger than normal and the components deliberately designed to last for about twenty years with almost no maintenance. The instruction manual would enable the users to carry out the maintenance themselves and learn by doing it.

ROAD/RAIL VEHICLE

In the mid-1950s Lucas Aerospace (Rotax) spent over £1 million developing an actuating mechanism whereby a set of pneumatic tyres could be brought down into position so that a railway coach could run on the roads. In its railway mode, a metal rim still ran on a metal track, which in practice resulted in all the shocks going up through the superstructure. Inevitably, this meant a large rigid superstructure of the type we have inherited from Victorian rolling-stock design.

But again, there is another approach, which was followed up by the Lucas workers and Richard Fletcher and his colleagues at the North East London Polytechnic. By using a small guide wheel with servomechanical feedback, the vehicle can be steered along the track with the pneumatic tyres running on the rails.

With the guide mechanism retracted, the vehicle can be used conventionally on the road. This provides the basis of a flexible lightweight vehicle which is capable of going up a rail incline of one in six.

Normal railway stock, because of the low friction between the metal rim and the metal track, is capable of going up an incline of about one in 80. This means that when a new railway line is being laid, for example, in the developing countries, it is necessary literally to flatten the mountains and fill up the valleys, or build tunnels and viaducts. Typically, this costs £1 million per track mile. With the hybrid vehicle, it is possible to follow the natural terrain and lay down new railway lines for £20,000 per track mile. The vehicle can of course be run on disused tracks to service remote areas.

A prototype of the road/rail vehicle has been built at the North East London Polytechnic and tested out on the East Kent railway line with great success. In parts of Britain, there is a growing interest in a vehicle of this kind since it could provide the basis for a truly integrated transport system with vehicles running through

Drawing up the Corporate Plan at Lucas Aerospace

our cities like coaches and then moving straight on to the railway network.[1]

KIDNEY MACHINES

The Lucas workers do not merely design and build *new* products. There are one or two existing products in Lucas Aerospace which they would like to see produced at a much greater rate. One of those is the home dialysis or kidney machine. In the mid-seventies, the company attempted to sell off its kidney-machine division to an international company operating from Switzerland. We were able to prevent them doing so at that time by both threats of action and the involvement of some MPs. When we checked on the requirements for kidney machines in Britain we were horrified to learn that 3000 people die each year because they cannot get a machine. In the Birmingham area, if you are under fifteen or over forty-five you are allowed, as a medical practitioner put it so nicely, 'to go into decline'. The doctors sit like judge and juries with the governors of hospitals deciding who will be saved. One doctor told us how distressed he was by this situation and admitted that sometimes he did not tell the families of the patients that this was happening as it would be too upsetting for them.

We were disgusted when we saw, in an ITV programme, an interview with a teacher who was over forty-five and being allowed to 'go into decline'. She said she was going to commit suicide at some stage so that her grandchildren would not see her going through the progressive stages of debility. Ernie Scarbrow, the secretary of the Combine Committee, said: 'It is outrageous that our members in Lucas Aerospace are being made redundant when the state has to find them £40 a week to do nothing except suffer the degradation of the dole queue. In fact the £40 a week amounts to about £70 a week when the cost of administration is taken into account. Our workers should be given this money and allowed to produce socially useful products such as the kidney machine. Indeed, if the social contract had any meaning and if there were such a thing as a social wage, surely this is the kind of thing which it should imply, namely having forgone wage increases in order that we could expand medical services, we should then have the opportunity of producing medical equipment the community requires.'

TELECHIRIC DEVICES

One of the most important political and technological proposals in the Corporate Plan is for the design of 'telechiric' (hands at a distance) devices. With these systems, the human being would be in control, real-time, all the time, and the system would merely mimic human activity, but not objectivise it. Thus the producer would dominate production, and the skill and ingenuity of the worker would be central to the activity and would continue to grow and develop. This would link human intelligence with advanced technology, and help to reverse the historical tendency to objectify human knowledge and thereby confront the worker with an alien and opposite force, as described earlier.

The methods by which the Lucas workers arrived at the concept of this product range are in themselves revealing. It was suggested to them that it would be highly socially responsible if a means could be found of protecting maintenance workers on North Sea oil pipelines. These maintenance workers experience a very high accident rate because of the depths at which they have to work.

Since they had been conditioned by traditional design methods, they immediately thought of a robotic device which would eliminate the human being completely. However, as they began to consider the programming problems of getting a system which would recognise which way a hexagon nut was about (let alone if it had a barnacle on it!) and select the correct spanner and apply the correct torque, they recognised what a difficult task this would be. Yet it is the kind of task that skilled workers can perform even 'without thinking about it'. They simply have to look at the diameter of a nut and bolt, and will know through years of experience what torque they can apply to it without wringing it off and yet at the same time tightening it sufficiently so that it won't become loose again. Without any 'scientific knowledge', such as the torsional rigidity of the bolt or the shear strength of its material, they will get it right repeatedly.

That is to say, workers don't express this knowledge in writing or speech. They demonstrate their knowledge and intelligence in what they do. The common sense and tacit knowledge described in Chapter 4 would be pivotal for systems of this kind.

Comparing the levels of intelligence of robotic equipment of that kind with total human information-processing capability, we have seen that the order of things is the machine 10^3 to 10^4 and the human being 10^{14}. This 10^{14} brings with it, however, consciousness, will, imagination, ideology, political aspirations; and these are precisely the attributes which employers regard as disruptive.

PEOPLE ARE TROUBLE – MACHINES OBEY

One doesn't have to engage in sociological research to work this out. The multinationals and the employers are so arrogant that they have constantly put it in writing in case we might fail to understand. A decade ago the *Engineer* had a headline which said, 'People are trouble, but machines obey.'[2] It is, therefore, no accident that systems today are designed around the trival 10^3 whereas the 10^{14} is deliberately suppressed. It is a political act which reflects the power relationships at the point of production.

The Lucas workers feel that there are hazardous, dangerous jobs which should be automated out of existence. What they were questioning was the politics of elevating these design methods to universal principles.

SILENCING THE WORKER

It has been pointed out that technological change viewed thus has more to do with the exercise of control over the workforce than it has to do with increasing productivity.[3] Andrew Ure, in his *Philosophy of Manufactures* puts it even more clearly when he says that

> the industrialists aim to take any process which requires peculiar dexterity and steadiness of hand, from the cunning workman, and put it in charge of a mechanism so self-regulating, that a child may superintend it. The grand object therefore, of the manufacturer is, through the union of capital and science, to reduce the task of his workpeople to the exercise of vigilance and dexterity appropriate to a child.

Architect or Bee?

The extent to which capital and science have succeeded in achieving this was dramatically illustrated in the July 1979 issue of the *American Machinist*. It reported that an engineering firm had found that the ideal operators for its numerically controlled machining centre were mentally handicapped.

One of the workers held up as ideal for this type of work had the intelligence level of a twelve-year-old. The employer pointed out in gloating terms, 'He loads every table exactly the way he has been taught, watches the Moog operate and then unloads. It's the kind of tedious work that some nonhandicapped individual might have difficulty in coping with.'

This would have been laudable had the objective been to provide work for the mentally handicapped, but what happens in installations of this kind is that some of the most highly skilled, satisfying and creative work on the shop floor, such as turning and milling, is so deskilled by these new technologies that it is rendered suitable for twelve-year-olds.

This historical process of deskilling[4] is an important means through which the employer extends his control over his employees. But in a wider sense, it destroys the social and cultural values which surround the exercise of those skills, and the means by which they are acquired. Indeed, we seem to have seriously underestimated the educational, cultural and other significance of skill and craftsmanship.[5]

Thus the significance of raising these issues through the very specific proposals surrounding the telechiric devices. The Lucas workers are indeed developing profound political ideas; so also are those who, in the wider sense, are proposing human-centred systems even in the field of high-level intellectual activity like design.

SOCIAL INNOVATION

We believe it is arrogant for aerospace technologists to think that they should be defining what communities should have. The Lucas workers were deeply conscious that if the debate were limited to industrial workers of this kind, it would represent a new form of elitism. We therefore made strenuous efforts to involve wide sections of the community at large in discussions around these issues in order to interact with them and learn from them. We sought, through the local trade unions, political parties and other organisations in each area, to get people to define what they needed, and to begin to create a climate of public opinion where we could force the government and the company to act.

To this end, the Lucas workers cooperated with Richard Fletcher at the North East London Polytechnic to convert a coach into one of their hybrid road/rail vehicles. As a way of consciousness-raising the vehicle was used as a technological agitprop with a photographic exhibition, slides and videotapes describing the concepts underlying the Corporate Plan and showing some of the prototypes in action. Local trade-union branches, trades councils and community groups sponsored visits of the vehicle to different cities and these culminated in large public meetings where discussions took place between technological and industrial workers and members of the public.

Part of the exhibition in the vehicle was a display of photographs composed by Dennis Marshall, a skilled worker at Lucas Aerospace. It vividly demonstrates the way that the ideas embodied in the Corporate Plan have released not only the technical creativity, but also the artistic creativity among the employees. Dennis Marshall has produced beautiful and vivid depictions with his camera of pollution, decay of inner cities, neglect of railway systems, and nuclear hazards. When I used these as illustrations to one of my talks at the Royal College of Art, the people there were amazed that an industrial worker could have produced such impressive work. I suggested that we would all benefit if they came to work at Lucas and made way for Dennis Marshall at the RCA.

TRADE-UNION RESPONSE

At national level, the trade-union movement has given very little support and encouragement to the Corporate Plan, although there have been some positive developments. The TUC has, for example, produced a half-hour television programme dealing with the Plan and this has appeared on BBC2 as part of its trade-union training programme for shop stewards.

Architect or Bee?

The Transport and General Workers' Union came out with a statement indicating that its shop stewards throughout the country should press for corporate plans of this kind. In 1986, the T&GWU produced a major policy statement on the conversion of the arms industry to socially useful production, as a phased reduction in arms expenditure. This is now the official policy of that union.

At an international level, the interest has been enormous. In Sweden, for example, they have produced six half-hour radio programmes dealing exclusively with the Corporate Plan, and have made cassettes which are now being discussed in factories throughout Sweden. They have also produced a one-hour television programme and a paperback book in Swedish dealing with the Corporate Plan. Similar developments are taking place elsewhere. In Australia, there have been television and radio programmes, including the *Science Show*, dealing with the Lucas Plan ideas. The Metalworkers' Union has produced a number of reports on the possibility of using resources such as railway workshops for the development of new forms of transport. The Australian government has also established a Commission for the Future, and I was the speaker at one of the conferences that launched it.

In the past, our society has been very good at technical invention but very slow at social innovation. We have made great strides technologically, but our social organisations are virtually the same as several hundred years ago. One of the Swedish television interviewers said, 'When one looks at Britain in the past, it has been great at scientific and technological invention and frequently has not really developed or exploited that. The Lucas Workers' Corporate Plan shows a great social invention, but it probably is also the case that they will not develop or extend that in Britain. If this were true it would be very sad indeed.'

THE TECHNOLOGICAL ASPECT OF THE PLAN

Although the social and political implications of the Lucas Aerospace workers' campaign have received considerable attention, the technology contained within the Corporate Plan has largely been ignored, even though the workers themselves, in their Plan, placed considerable emphasis on the forms taken by the technology, the products and the manner of producing them. This is particularly true of criticisms of the Plan.[6]

This reluctance to deal with the technology of the Plan is on the one hand due to the remarkable incompetence of those on the Left in the field of science and technology, and on the other to an indifference to it, because it is perceived, as described earlier, to be 'neutral'.

The Lucas workers had sought to find, and debated in considerable detail, forms of technology that would give full vent to the creativity of the hands and minds of the workers, and that could be carried out through nonhierarchical forms of industrial organisation. For the fact is that for those at the point of production, the considerations of the technology, the design methodologies, and the nature of the labour process which arises from them, are of equal importance to the political considerations precisely because these workers do not separate one from the other. Indeed, one of the most positive features the Lucas workers saw arising from their Corporate Plan was the discussions which eventually took place with shop stewards and representatives of workers at all levels in industry, from scientists to semiskilled workers, in a range of companies from Vickers, Parsons, Rolls-Royce, Chrysler and Dunlop to Thorn EMI. These discussions centred not just around the political aspects, but gave rise to a profound questioning of the nature of the technology itself and the design methodologies used.

In the course of designing and building prototypes they discovered that, as one worker expressed it, 'Management is not a skill or a craft or a profession but a command relationship; a sort of bad habit inherited from the army and the church.'

They did not mean by this that forward planning, project management and the coordinating and synchronising aspects of a project are unimportant. They were suggesting, rather, that these conceptual and planning aspects of work should be integral to the labour process, thereby ensuring that those who do work also plan and manage it.

Historically, when the great master builders designed, planned, managed and built their own structures, there was, of course, a hierarchy, but it was based on a legitimacy of leadership in that those who were managing knew what they were talking about and were capable of exercising the skills themselves.

The objection expressed above is to that form of management that seeks to remove the internal conceptual part of work and place it in the hands of those who represent capital, capital which has now become external to the productive process. There are hordes of accountants, financial planners, monitors and other nonproductive workers who are simply there to act as police people for external capital. This is part of that wider process in which finance capital increasingly dominates industrial capital, a moribund stage in which the production of capital becomes more important than production itself.

None of this is to imply that there are not important project-planning, financial-control and other skills. What the Lucas workers were suggesting is that facilities should be provided for industrial workers to acquire these skills rather than to have them used against them in a crude set of power relationships. They have also shown, if only in embryo, that the design methodology used in a 'socialist technology' would have to be radically different from that which applies in our current technology.

At present, in the technologically advanced nations, highly qualified designers and technologists spend months drawing, stressing and analysing a prototype before telling the workers on the shop floor what should be done. These design stages involve rarefied, complex mathematical procedures which are necessary only because, for commercial reasons, materials have to be exploited to the full. Both the materials and the systems of the products are designed just to perform a precisely defined function for a very short length of time before the product is rendered redundant (planned obsolescence). The rarefied mathematical procedures are outside the experience of the mass of industrial workers and are used as a means of silencing their common sense.

DRAMATIC EXAMPLE

There is a tendency for computer specialists to imply that they have the solutions to all our problems without necessarily having much real design experience behind them. Dramatic examples of what can result from this are already coming from the United States. At one aircraft company they engaged a team of four mathematicians, all of Ph.D. level, to define, in a programme, a method of drawing the afterburner of a large jet engine. This was an extremely complex shape, which they attempted to define by using Coons' patch surface definitions. They spent some two years dealing with this problem and could not find a satisfactory solution.

When, however, they went to the experimental workshop of the aircraft factory, they found that a skilled sheet-metal worker, together with a draughtsman, had actually succeeded in drawing and making one of these. One of the mathematicians observed, 'They may have succeeded in making it but they didn't understand how they did it.' This seems to me to be a rather remarkable concept of reality. It dramatically illustrates the manner in which the three-dimensional skill of draughtsmen and skilled workers can be thoughtlessly eliminated in this drive to replace people by equipment. All their knowledge of the physical world about them, acquired through years of making things and seeing them break and rupture, is regarded as insignificant, irrelevant or even dangerous.

With the prototypes developed for the products proposed in the Corporate Plan, the methodology of production was quite the reverse of the above. Workers on the shop floor had every opportunity of giving full vent to their skills and creativity since the prototypes were designed more by tacit knowledge than by analysis.

It is a sad reflection of the specific form technology takes in this society that this wealth of knowledge is deliberately eliminated. Clearly, any talk of industrial democracy with this kind of technology is simply a deception.

AN ELEMENT IN DESIGN

It will undoubtedly be argued by the authoritarians, both of the Right and of the Left, that the Lucas workers' approach to technology is romantic, unrigorous and unscientific. Such a view ignores the fact that a desire to meet real social needs is a vitally important stimulus to good quality and creative design, and is a qualitative element of design which cannot be treated in a mathematical way as the quantitative elements can.

Nor are the Lucas workers alone in taking this view of science and technology. In a recent paper, one of the country's leading technologists and academics, Howard Rosenbrock, had this to say:

> My own conclusion is that engineering is an art rather than a science and by saying this I imply a higher, not a lower status. Scientific knowledge and mathematical analysis enter into engineering in an indispensible way and their role will continually increase. But engineering also contains elements of experience and judgement and regards all social considerations and the most effective ways of using human labour. These partly embody knowledge which has not yet been reduced to exact and mathematical form. They also embody value judgements which are not amenable to the scientific method.[7]

HUMAN-CENTRED SYSTEMS

Howard Rosenbrock, a highly creative scientist, has developed advanced interactive graphic systems which actually place the designer and human intelligence at the centre of the design process.

In the field of computer-aided design, he cautions against the computer becoming an automated design manual, leaving only minor choices to the design engineer. The automated design manual approach, he says, 'seems to me to represent a loss of nerve, a loss of belief in human abilities and further unthinking application of the doctrine of the Division of Labour.' The designer is reduced to making a series of routine choices between fixed alternatives, in which case 'his skill as a designer is not used, and decays'.[8]

Rosenbrock's interactive graphic systems are described in his book together with their basic mathematical techniques.

He has developed graphic displays from which the designer can assess stability, speed of response, sensitivity to disturbance and other properties of the system.[9] This he and his colleagues did by using the inverse Nyquist array. Having demonstrated through his

Drawing up the Corporate Plan at Lucas Aerospace

computer-aided design system that alternative approaches can be applied to problems of this kind, Rosenbrock then raises the much wider question as to whether we are not cutting off options in other fields of intellectual work in rather the same way as we have done at an earlier historical stage in the field of manual work. He has termed this the Lushai Hills effect.

Other computer scientists, J. Weizenbaum among them, are now seriously questioning where their work is taking society, and what its impact on human beings and their self image is likely to be.[10]

What is often lacking in honest expressions of concern of this kind is an economic and political analysis of the forces in society which control and distort science and technology to fulfil specific class roles.

Thus the discussions at Lucas and elsewhere should really be viewed in the context of the overall questioning of the way science and technology are developing under advanced capitalist society. It is linked with the wider challenges workers are attempting to make against the way technology is being organised: the Green Bans Movement in Australia, the attempts by Fiat workers in Italy to transcend the narrow economism which characterised trade-union activity for so long, and the courageous struggle by the women at Algots Nord in Sweden.[11] The Green Bans Movement was started by the building labourers in Australia in the seventies. It was an attempt to link their industrial strength and strike power to community groups and conservationists, in order to prevent developers from destroying buildings of architectural, cultural or social significance. The bans also applied to areas of land held to be important to the local community and regarded as part of a heritage which should be protected for future generations. The Fiat workers proposed alternative products. The women workers at Algots Nord, when faced with the closure of their clothes factory, took it over and went out and asked the community, and occupational groups like nurses and electricians, what kind of clothes or working garments *they* would like to have, and worked with them to design new product ranges.

All these forces can be linked together in a challenge to the system as a whole, and as a forerunner to a transformation of society which will take it away from its present exploitative, hierarchical form to a new type of society, which, the founder of cybernetics, Norbert Wiener, once said:

> differs from those propounded by many fascist successful businessmen and politicians. People of this type prefer an organisation where all information emanates from the top and where there is no feedback. The subordinates are degraded to become effectors of an alleged higher organism. It is easier to set in motion a galley or factory in which human beings are used to a minor part of their full capacity only, rather than create a world in which these human beings may fully develop. Those striving for power believe that a mechanised concept of human beings constitutes a simple way of realising their aspirations to power. It maintains that this easy way to power not only destroys all ethical values in human beings, but also our very slight aspirations for the continued existence of mankind.

The new technologies highlight the fact that we are at a unique historical turning point.

Eight
THE LUCAS PLAN – TEN YEARS ON

On the front page of the now famous Lucas Workers' Plan for Socially Useful Production there is the statement that 'there cannot be islands of social responsibility in a sea of depravity'. Lucas workers themselves never believed that it would be possible to establish in Lucas Aerospace alone the right to produce socially useful products. Only some of their baton wavers seemed to believe that, or those who sought to decry their activities by suggesting they were utopian.

What the Lucas workers did was to embark on an exemplary project which would inflame the imagination of others. To do so, they realised that it was necessary to demonstrate in a very practical and direct way the creative power of 'ordinary people'. Further, their manner of doing it had to confirm for 'ordinary people' that they too had the capacity to change their situation, that they are not the objects of history but rather the subjects, capable of building their own futures.

The Lucas workers could see about them the grotesque absurdities of modern industrial society. They were aware of the growing powerlessness and frustration of masses of people as decisions were concentrated in the hands of vast multinational corporations whose size and activities dwarfed that of nation states. It was a courageous attempt to repossess that precious ground of decision-making which planners, managers and coordinators were removing from them. They highlighted and built on the major contradictions in industrial society.

The audit of their own skills and abilities, and the surveys in different factories and workshops analysing and assessing the production equipment, product ranges and skills, represented an enormous extension of consciousness, since we are all of us conditioned to view the world from the one lathe we operate or the one desk from which we function. Never are we encouraged or allowed to take a panoramic view of our industry and see how that fits into a wider pattern of society.

Prototypes of the products included in the plan were built and displayed on a wide scale. Lucas workers then went to the Labour government and pointed out that its manifesto had said that it stood for 'an irreversible shift of power in the interest of working people'. The government, and in particular the Department of Industry, was clearly bewildered by the notion that one might think of products for their use value rather than their exchange value. In addition, most trade-union bureaucracies bitterly resented a rank-and-file activism which they perceived as a challenge to their leadership. The idea that leadership might be catalytic, enabling and supportive was beyond them.

The whole exercise demonstrated in a very direct way to the Lucas workers the nature of power relationships. For example, when they proposed the manufacture of heat pumps using natural gas in internal combustion engines, the company turned the proposal down, saying that it would not be profitable and was incompatible with their product range. Burnley workers subsequently revealed that the company had had a report prepared for them by American consultants showing that the market for the heat pump would have been some £1000 million in the private and industrial sectors in the EEC countries by the late 1980s. Lucas would be willing to forgo a market of that kind to demonstrate that it, and it alone, would decide what was made, how it was made, and in whose interests it was made. Lucas workers then quickly recognised that they were dealing not just with an economic system but with a political system concerned with the retention of power.

When the company moved on to the offensive and victimised some of the leading stewards, amid worldwide protest, there was inadequate support from union leaderships. The Lucas Plan was turned down by the Labour government and rejected by trade unions, with the exception of the T&GWU and some peripheral support from ASTMS. Lucas workers then felt that the key strategic position was to diffuse the ideas as widely as possible through the Labour and trade-union movement. They formed a

Trade Union combine committee and produced a number of very well worthwhile reports. Increasingly they also entered into discussions with those who were contesting elections in local government.

THE GREATER LONDON ENTERPRISE BOARD

In its manifesto for the 1981 election, the Labour Party in London committed itself to restructuring industry along the lines of the Lucas Workers' Plan if it came to power.

Once elected, Labour sought enthusiastically to meet that commitment. They set up an Industry and Employment Branch at County Hall, and an early part of its work was to establish the Greater London Enterprise Board. The GLC provided the board with some £30 million a year and in its first two and a half years of operation it established, restructured or assisted 208 companies and created directly some 4000 jobs, with many more jobs created indirectly. Based on its direct investment in industries, it was able to create jobs at some £4700 per job, whereas if somebody is unemployed in Britain and has a couple of dependants, it costs the taxpayer £7000 a year.

It was only to be expected that a project as innovative as this – one for which there were no real precedents – would encounter many difficulties. Apart from the problems arising from its innovative nature, there was also a conflict between long and short-term investment policies. Frequently, the design, research and development of new products requires ten or even fifteen years. However, such products then become part of the nation's productive activity, and help to create real wealth. There is growing concern, even in City circles, that the City is concerned only with those investments which will show a quick return in the short term.

GLEB's achievements, limited though they are, have to be seen in the hostile context in which GLEB found itself at both a national and local level. Certainly, it was seeking to swim against the economic tide. It came into existence and is attempting to operate at a historical stage when finance capital is dominating industrial capital, and when there is a dramatic decline of the British manufacturing base. In the Greater London area, the trend, stimulated and supported by the Thatcher government, has been to transform London into the finance centre of the world, with precious little support for its manufacturing base.

To all these difficulties must be added the government's determination to abolish the Greater London Council. Of particular significance were the months of uncertainty in 1985 surrounding the GLC's future and then its final abolition in March 1986. This did much to damage GLEB and its projects. And the decision of the government in March 1986 to refuse to release the remaining £8 million which the GLC had provided for GLEB will inevitably mean that for the foreseeable future its capacity to act will be greatly constrained and the future of many of its projects precarious.

However, what is clear is that many of its exemplary projects indicate alternatives for the future – its policies on cooperative enterprises, its framework for enterprise planning and equal opportunities, and the relationship of these to the overall London industrial strategy and enterprise planning.[1] GLEB's technology policy, in particular, has attracted both national and international attention and, indeed, emulation. That policy was based on an imaginative framework which was agreed with the GLC in early 1983.[2] This framework provided for the establishment of new and high-tech companies and it also provided for human-centred means of production. Perhaps its most innovative aspect was the proposal to set up a range of technology networks throughout London.[3] These would draw on the ideas of the Lucas Workers' Plan and also on the experiences of the science shops in Holland, the innovation centres in West Germany and the more desirable aspects of science and technology parks.

GLEB'S TECHNOLOGY NETWORKS

A pivotal aspect of the policy was that of linking two of London's great resources – the skill and ingenuity of the people of London and the facilities of London's higher educational infrastructure in seven polytechnics, three universities and many teaching hospitals and colleges.

Two basic forms of network were proposed and established:

geographically based networks and product-based networks. A north and east network and a south and east network provide facilities and support for the communities in their respective areas. In January 1987, four of the west London boroughs, together with GLEB, put forward a proposal for the establishment of a west London network, calling on each of the west London boroughs to provide some funding, and on GLEB to provide some of the technical expertise required in setting up and running it.

These geographical networks have proved more difficult to establish than the product-based networks. Firstly, what constitutes a community? Secondly, when meetings of the 'community' were called to define the types of products and services they wished to have from the networks, these meetings were invariably attended by the habitual meeting-goers who tend to turn up at all such gatherings, whether they are about the bomb, unemployment or whatever, and who tend to say the same things at each meeting. To get through to real people proved to be much more difficult. Further, many of the 'community activists' came from that tradition which is obsessed with the contradictions of distribution at the expense of a serious analysis of the contradictions of production. They often tend to be 'takers' rather than 'makers' and were sometimes more concerned about the control of resources than the creation of new ones.

However, the needs of those who are dying of hypothermia, or unable to get a kidney machine, or immobilised because they lack the resources to overcome a comparatively simple disability, are so clear, so obvious and so well defined that an organisation professing to deal with the requirements of the community cannot easily avoid providing these services. A structure gradually evolved that enabled the networks to respond creatively to community and other needs.

The product-based networks in general found it easier to establish themselves. They had the cohesion of a clearer framework. Three such networks were established: the New Technology Network, the Energy Network and the Transport Network. Each network is a combination of people, skills and physical facilities. The main physical facility is a workshop located close to a university or polytechnic but never actually on the campus. A university campus can be very alienating for the unemployed, for those whose experience of the world is real and experiential, for women's groups, ethnic minorities and the disabled. (They are probably also somewhat alienating for the students, but they don't have much choice at this stage.) So the buildings were located so that academics and the community could meet on neutral ground and the facilities and skills in the polytechnics could be utilised in a supportive fashion. In this way, academics who wish to support the communities in which they are based have a framework in which they can do this without being reduced to industrial fodder for the multinational companies. Furthermore, it is possible to provide exciting student projects rather than artificially contrived ones which frequently are highly demotivating for the students.

Each technology network is constituted as a company limited by guarantee with which GLEB has an annual funding agreement against set objectives and projects. The management committee of a network will typically involve representatives of local councils, trade unions, special interest groups and local academic institutions.

Even for high-tech projects, the policy is to provide a practical environment based on design by doing. Each of the networks has around six or eight technicians, engineers and support staff who appreciate the tacit knowledge of ordinary people and can relate to it. Each network building typically has four times as much workshop space as office space so that it is action-oriented, since there is a real danger that anything relating to local government can quickly degenerate into report-writing, and there are those who believe that a report constitutes a final product rather than being a guide to some real action which will follow. The issue still is to change the world and not just analyse it.

Each of the networks has tended to develop its own contacts and product ranges.

THE PROJECTS

The North and East Network has now been transformed into the London Innovation Network, and still serves the same geographical area while providing an innovation service for London as a whole. One of its most exciting features has been to see disabled people working with engineers and technologists on the design of new equipment for themselves. Some of the equipment is therapeutic and is related to new programmes of therapy supported by the local authorities. Some provides systems of mobility for the disabled and the elderly.

The energy network has produced an imaginative series of products and services. These range from energy audit systems (which analyse the energy requirements and costs of a building and suggest more effective ways of providing the heating necessary) to proposals for building conversions. Council tenants, short of funds, have been put in contact with building societies and banks who will provide forward funding so that buildings can be converted in an energy-conscious fashion and the tenants can then pay for this over a two-year period, at the end of which they will have a permanent gain as a result of reduced energy costs. Integrating dehumidifiers with energy-conscious systems reduces not only the heating costs but also the problems associated with condensation.

At the other end of the spectrum, expert systems are being developed, using very advanced computing techniques, in conjunction with some of the teaching hospitals. These systems provide the technology through which advanced expert knowledge can be diffused back into general practice and the community, thereby democratising decision-making between the general practitioner and the medical consultant. The data base is structured in a way that provides the medical practitioner and the patient with different treatment options, thereby encouraging a dialogue between them.

Such a form of expert system avoids the problems associated with many of those in the United States, in which the knowledge is only visible to a small elite of medical consultants. The medical profession downstream is deskilled. The London system means that people need not go into large, alienating, factorylike hospitals to be treated for each major illness, but may in future be treated in their local community by their own GP.

In the transport field, new forms of power-assisted bicycle have been developed. These are ecologically desirable forms of transport and will give people exercise more naturally than sitting on absurd exercise bicycles in their front room, going nowhere. The power-assisted bicycles would enable older people, who have given up cycling, to resume, since the power will assist them on inclines. But this would require an infrastructure of cycleways in cities, and the GLC was embarking on this before its abolition.

Other products include 'hush kits' to reduce the noise levels from vehicles in the inner-city areas, and these might provide the basis for future legislation.

Still in the field of transport, one outcome of the work on developing the Lucas workers' road/rail vehicle has been a new kind of composite tyre which will retain the desirable characteristics of a pneumatic tyre, but cannot be punctured.[4]

Polytechnics have supported the development of cell-immobilisation techniques. These make possible the production and storage of real ale, and have other applications, possibly in the field of yoghurt production.

Many of the networks are now beginning to provide new products as a way of opening up new markets, and they can suggest new products to existing GLEB and other London companies which will experience difficulties with their product ranges during coming years. Desirable products with a future will also become available, for example, to black businesses, which too often are constrained to ghetto trading, and to cooperatives, so that they are not left with the dregs of the market economy. As a result of all these activities, a product bank has now been built up containing some 1500 products at various stages of development, from the idea or concept to prototypes and items in production.

The product bank is exciting, especially the way it has been developed. Special-interest groups concerned about energy conservation have been able to develop product ranges. The disabled have shown great creativity not only in thinking up alternative products for themselves, but in designing and, in many cases, making them. The networks have also caused cultural shifts. Frequently, ideas are regarded in our universities and polytechnics as important only at the conceptual level, and it is sometimes held to be second-rate or even slightly obscene to be involved in production. Both the academics and their students have enjoyed access to a socially responsible framework in which to diffuse scientific and technical ideas through society.

The teaching hospitals frequently develop individual pieces of equipment for research or patient-specific treatment. It is often possible to elevate these to product ranges to make specialist treatment more widely available and at the same time provide work for those who make the new products.

All the products are geared to opening up new social markets, and this, linked with popular planning, could provide a structure in which we would have much of the dynamism of the market without its indifference to environmental impacts and the real needs of human beings.

NEW FORMS OF TECHNOLOGY – THE ESPRIT PROJECT

There has, in recent years, been a growing tendency to assume that there is only one form of technology – that which we may now think of as 'American technology'. This view constitutes a kind of Taylorism at the macro level, a belief in the notion of the 'one best way'. A richer and more sensitive way to view technology would be to perceive it as a cultural product, and since culture has produced different languages, different music and different literature, why should it not produce different forms of technology, forms which reflect the cultural, historical, economic and ideological aspirations of the society which will use them? Should there not be a form of European technology reflecting European aspirations (if more in the rhetoric than the reality) of motivation, self-activism, dignity of the individual, concern for quality etc., and reflecting also the reality of the European manufacturing base which is composed predominantly of medium-sized and small-scale units?

To explore the potential of this human-centred approach to advanced technology and to demonstrate its feasibility in the real world, ten partners from three European countries have come together within ESPRIT project 1217. The partners in the £3.8 million EEC ESPRIT project are from Denmark, Germany and the UK. Extending the Lucas proposals for telechiric devices, it will set up a demonstration centre in London in 1988 and display, for the first time ever, a human-centred computer-integrated manufacturing (CIM) system. It will shift the ground from theoretical discussions to a practical demonstration of the potential for the integration of advanced computing systems with human skill and ingenuity.

The CIM system will provide a complete manufacturing capability right through the spectrum from computer-aided design (CAD). This area of work will be undertaken by the Danish partners, drawing on systems-design work developed at UMIST. They will use a novel capability in which the designer can really sketch and those on the shop floor can converse with the designers and express their ideas through sketches, thereby creating a dialogue between the shop floor and the design office to the enhancement of both areas.

From the design area, the work will flow through its production scheduling and sequencing via a computer-aided production (CAP) system which will be developed by the German partners. They are exploring exciting possibilities of rendering visible, via the computers, the production sequencing, and are considering linking this to innovative organisational forms which include 'islands of production'. The actual machine tools and the production cells will be based on computer-aided manufacturing (CAM) systems which will be designed and developed in the United Kingdom. The CAM system will build on the skill of craft workers and enhance that skill so that they in fact become production cell managers.

This human-centred concept will, we hope, dramatically shift the paradigms of systems design. The complexity of such systems at the design level will be a challenge to the ingenuity of a new multidisciplinary design team which will draw, not merely on technical and scientific expertise, but also on expertise in the fields of social science, psychology and political science. It is also likely to have an important impact on industrial relations. The last ten years have been characterised by trade unions simply reacting to the forms of technology imposed on them by the large corporations. The human-centred CIM system could mean that trade unions need no longer be caught in Wapping-like situations but could demand alternative forms of systems that meet the requirements of their members and, in the long term, demand technologies that enhance human skill rather than marginalising it.

HUMAN-CENTRED CONCEPT

The human-centred concept is based on the premise that a computer-integrated manufacturing system will be more efficient, more economical, more robust and more flexible if designed to be run by a human than a comparable unmanned system. The operator, who is really a cell manager, will run the cell with the aid of powerful software tools. His or her job will include the following tasks:

Creation of the machine programs, from 'part data' originated elsewhere, by using high-level software tools. The colour-graphics systems available today are an example of these tools.

Optimisation of these programs, using the operator's skills and experience to minimise the cutting time.

Machine scheduling of the cell's job list to make the most savings possible; for example, by running similar jobs sequentially with the same tool set-up.

Programming the work handler to load and unload all the parts that can be gripped by standard grippers, using powerful software tools with a simulation facility.

Doing all the jobs needed to run the cell, for example, load parts that cannot be handled by the robot work handler, change tools and de-burr parts.

The human-centred system will be more efficient than conventional fully automated systems because the operator can use his skills and experience, with the aid of powerful software tools, to optimise the machining programs and the job scheduling in the cell. It will be more flexible because any job that the machines can cope with can be machined in batch sizes of one upwards. It will be more robust because there is much less dedicated automation and electromechanical complexity, so that when a failure occurs the cell may be instantly reconfigured to allow for greater human intervention, and the fault will take less time to fix because there are fewer dedicated subsystems. It will be more economical because it is designed to be more efficient, more flexible, have a higher uptime, lower running costs, cost less to buy and take less time to commission.

ESPRIT PROJECT 1217 (1199)
HUMAN-CENTRED CIM SYSTEMS

A manufacturing cell comprising integrated CAM, CAD and CAP modules will be developed and installed at a user site. In the first phase the CAM cell will consist of two lathes and a work handler; a prismatic machine will be added in the second phase. The CAD system will integrate a drawing board with a CAD work station and the CAP system will be designed especially for shop-floor operation.

The CAD section will be handled in Denmark, the CAP in Germany and the CAM in Britain, with the Greater London Enterprise Board acting as prime contractor. In addition to these three areas, the University of Bremen will examine the educational requirements for such enhanced systems.

ROLE OF THE OPERATOR IN
A HUMAN-CENTRED CIM CELL

The operator's role includes some or all of the following tasks, depending on the cell configuration:[5]

Produce and optimise part programs and work-handler load/unload programs with extensive use of simulation tools.

Use tool-wear data to determine when tool tips should be changed.

Reduce machine set-up times, e.g., schedule similar jobs sequentially.

Optimise machine utilisation, e.g., make sure there is no more than one machine awaiting attention at a time.

Quickly switch to high-priority jobs.

Allocate work to machines in the cell.

Feedback problems with production of the part to the designer.

Prove-out new part programs and load/unload programs.

Decide when to have tea breaks while still keeping the cell running.

Manual recovery from problems, e.g., tool breakage.

Any nonautomated activity, e.g., tool changing, loading/unloading awkward parts that the work handler cannot grasp.

Inspect the parts produced and carry out any consequent corrections to the machining process.

Devise and carry out de-burring strategies and systems and advise on appropriate equipment.

INTERACTION OF CAD, CAP AND CAM ELEMENTS

The CAM cell operator has considerable responsibilities above those that have been traditionally accepted.

The management of this change in the balance of power between the office and shop floor will be critical to the success of the concept.

CAD design is definitely not a shop-floor operation, but, despite the advent of modern technological aids, designers do not have the capability for developing competent part programs.

Communications protocols between CAP, CAD and CAM systems and the elements within these systems are largely human-independent, and current standards initiatives such as MAP and IGES are thus likely to be compatible.

The following are examples of interactions between the three elements:

CAD to CAM: the part geometry, material and blank geometry will be down-loaded, together with questions like 'Is this design OK for machining?'

CAM to CAD: request changes to the design to suit the machine and operating questions such as 'May I use a different material?'

CAP to CAM: job schedules with priorities, urgent jobs, warnings of changes to schedule, estimated-time requests for new designs.

CAM to CAP: real-time information on status of machine and accomplishments, estimates of times for new jobs, completion estimates for urgent work.

BENEFITS OF HUMAN-CENTRED SYSTEMS

The economic benefits will stem mainly from the increased efficiency achieved by incorporating the skills and experience of the operator into the running of the cell.

Human-centred systems will provide more stimulating and challenging work, resulting in a higher degree of motivation. They will require greater intelligence, involvement and commitment from the operator.

The human-centred concept is well suited to European industry, where workers have a high degree of skill and now have many years' experience with computer control of machines.

LIST OF CRITERIA

A prime objective of the three-year ESPRIT project that we have embarked on is to establish criteria for the design of human-centred systems; this list is only a summary of the present state of the art.

Technology can evolve in a direction that includes the skill and initiative of workers, who make it more productive and evolve new skills appropriate to the new kinds of technology.

The shop-floor workers have knowledge of the production process and especially deviations from the normal, so detailed specifications of the machining processes should be made on the shop floor.

At this stage design will only be undertaken in the office, but the design information received at the shop floor could be returned, with suggestions for improvement relating to production, to the office.

All manufacturing data should wherever possible be manipulated in a high-level form appropriate to the users' interactions and not at the machine-level program; appropriate software tools must be developed to aid this.

CAD technology can be adapted to encompass tacit knowledge built up from experience and to some extent tied to the use of the drawing board by developing a system concept combining the use of the CAD system and the drawing board.

The human-centred operator-interface software packages would incorporate flexibility to encourage the development of alternative ways of working.

CAP systems would provide well-structured information about the current and expected state of the manufacturing process to any employee.

Real-time monitoring will be tailored to the planning and control functions to eliminate any time lags in the CAP system.

The operator's ability to deal with any unforeseen circumstances and cure malfunctions should be enhanced by structuring the information-processing systems for human operation.

Defined and compatible linkages are required between the technological concepts and the skill and experience of the workers. This will be accomplished by off-line training and on-line reinforcement.

A practical exploration of the potential of human-centred manufacturing systems is long overdue. The frantic linear drive forward towards a workerless factory may well prove to be a solution lacking systems robustness and flexibility and even then applicable only to very narrow segments of industry which are in no way typical of the mass of European industry. Human-centred CIMs may also provide one important element in a wider strategy to cope with structural unemployment and deskilling and begin to address trade-union concerns for the quality of working life and the humanisation of work.[6]

SOCIALLY USEFUL PRODUCTION

I have referred frequently to the desirability of socially useful production as an alternative to growing structural unemployment. What then is socially useful production? And what criteria would we look for in a socially useful product?

Interestingly enough, the Lucas workers never set out to define socially useful production in an academic way, but rather counterposed it as an alternative to forms which they regarded as obviously not socially useful, for example, large-scale systems of mass destruction. Wittgenstein once said something to the effect that words define themselves by their use, and that has tended to be the case with socially useful production.

Given that the Lucas workers identified 150 products and services which they could provide, and subsequently the Technology Networks developed hundreds of socially useful products, we can begin to construct a tentative list of those attributes, characteristics and criteria which constitute socially useful production. It is not suggested that all these will be present in any particular socially useful product or production programme, but rather that some of these are key elements within it.

1. The process by which the product is identified and designed is itself an important part of the total process.
2. The means by which it is produced, used and repaired should be nonalienating.
3. The nature of the product should be such as to render it as visible and understandable as is possible and compatible with its performance requirements.
4. The product should be designed in such a way as to make it repairable.
5. The process of manufacture, use and repair should be such as to conserve energy and materials.

6. The manufacturing process, the manner in which the product is used and the form of its repair and final disposal should be ecologically desirable and sustainable.
7. Products should be considered for their long-term characteristics rather than short-term ones.
8. The nature of the products and their means of production should be such as to help and liberate human beings rather than constrain, control and physically or mentally damage them.
9. The production should assist cooperation between people as producers and consumers, and between nation states, rather than induce primitive competition.
10. Simple, safe, robust design should be regarded as a virtue rather than complex 'brittle' systems.
11. The product and processes should be such that they can be controlled by human beings rather than the reverse.
12. The product and processes should be regarded as important more in respect of their use value than their exchange value.
13. The products should be such as to assist minorities, disadvantaged groups and those materially and otherwise deprived.
14. Products for the Third World which provide for mutually nonexploitative relationships with the developed countries are to be advocated.
15. Products and processes should be regarded as part of culture, and as such meet the cultural, historical and other requirements of those who will build and use them.
16. In the manufacture of products, and in their use and repair, one should be concerned not merely with production, but with the reproduction of knowledge and competence.

This list is by no means exhaustive, and is being developed day by day by the Technology Networks and groups worldwide who are now putting the concept of socially useful production into practice. The examples above demonstrate the capacity of quite ordinary people to question the direction in which technology is going, and demonstrate in a practical way some of the alternatives, and the processes by which we develop those alternatives. As we set out to do so, there is a danger that our sense of what is necessary will be silenced by technocratic, scientific jargon. We should not permit this, nor should we be intimidated by the determinism of science and technology into believing that the future is already fixed.

LUCAS PLAN SIGNIFICANT

The Corporate Plan is significant in that it was a very concrete proposal put forward by a group of well-organised industrial workers who had shown in the past, by the products they had designed and built, that they were no daydreamers. It demonstrated clearly to a whole host of scientific and technical workers, through the medium of their own jobs, what the limits of the system are. Many of them actually used to believe that the only reason society didn't have nice, socially useful products was that nobody had thought of them. The fact that these products are being built and are still being rejected, both by the government and the company, demonstrates in very dramatic terms the kind of priorities dominant in this society.

Nine
SOME SOCIAL AND TECHNOLOGICAL PROJECTIONS

COMMUNITIES

There is, correctly, much discussion about the need to revitalise communities and establish local job-creation schemes. One possibility worth considering is that of community-owned enterprises on non-hierarchical lines, closely integrated with local needs for equipment and services. However a serious aspect of this 'community rejuvenation' is the drift towards what I shall call industrial feudalism. Our economy is now dominated by the massive multinational corporations and financial institutions. The role of even nation states is quite subordinate to these, since they set the economic, and increasingly the political, framework within which the governments of the individual nation states are allowed to legislate. These vast corporations spearhead the so-called technological revolution, and distort its development to meet their needs for profit maximisation. The underlying assumption is that of a rapacious economic system fired by ever increasing consumption and production. Apart from the waste of energy and materials their throwaway products bring in their wake (not to mention widespread pollution), they are becoming increasingly capital-intensive rather than labour-intensive. Throughout the technologically advanced nations, they are displacing millions of workers and are permanently destroying jobs and skills while extending their control over the cultural as well as the economic and social lives of the mass of the people.

In addition to this, the ecological crisis is likely to assume even greater political significance and it is important that it should not be seen as a middle-class preserve. After all, it is always the stream in which a worker does his bit of rough fishing that will be polluted, seldom if ever will it be the salmon stream in Scotland. It is usually the working-class community that will have a motorway running through it, not the stockbroker belt in Surrey. It is usually the working-class playground that will have filth belched out upon it from local factories, hardly ever the middle-class golf course. Workers do not stand to gain from the misuse of science and technology. They make no profit from the pollution of the rivers, the seas and the air. It is in their class interests to resist these things and it is vital that they should be involved.

Public hostility to the naked power of this 'democratic oppression' is growing, and ranges from the reluctance of young people to work for the big multinationals, to the more dramatic actions against businessmen in West Germany and Italy. These are as yet isolated indicators of what may become a very significant movement, when large sections of the labour and trade-union movement finally realise that we are dealing with structural unemployment rather than the cyclical forms we experienced in the past.

THE NEW FEUDALISM

The vast multinationals are increasingly conscious that there is going to be a backlash from society at large due to the way they are distorting its development. Some of my acquaintances, who are well-placed technocrats in these vast companies, tell me that they are about to engage in what they call 'programmes of enlightened self-interest', and are planning to move into the community and job-creation field.

The idea is that the big firms will supply some funds, and second their executives (with all the elitism that implies) to set up small-scale community enterprises. In this way they hope to be able to placate the public, on the one hand, and thereby, on the other, be left free to get on with the serious business of maximising their profits at the commanding heights of the economy.

Computerisation and automation will mean that smaller numbers will be required to run the large corporations. These will be a separate elite from the rest of the community and highly organised on the 'business union' basis as in the United States. They will be true 'Corporation Men', satisfied economically with company cars, company houses and company medicare schemes (not insignificant, with the systematic attempt to break up the National Health Service). There will be funds for the schooling of the Corporation Man's children, special superannuation schemes and of course expense accounts for overseas travel, entertaining and other corporation 'responsibilities'.

A large sector of those who remain unemployed will be left to fiddle around with 'community work'. These activities will be deliberately chosen because they yield no economic power. It will be a sort of therapeutic, do-it-yourself social service. The industrial feudal lords will sit in the multinational headquarters while the peasants scratch out a living in the deprived communities in which they reside. In practice, this will mean that they will spend their time repairing, cleaning up, modifying and recycling the rubbish which the large corporations are imposing on them. While it is conceivable that some of these jobs will be craft-based and thus provide an outlet for some initiative and self-activism, the significant reality will be that they have no economic power and no industrial muscle. So a considerable proportion of the population seems destined to have no 'real job' – no paid employment. Caring activities, leisure activities, unpaid social work and community work are not recognised as real jobs. The social multiplier effects of unemployment will be alienation, drug taking, suicides, interpersonal violence and general degradation, all of which are evident in abundance in our inner-city areas.

We will all be assured, of course, that this kind of work will be nonalienating and will enhance our self-reliance. Both of these are highly desirable in themselves, but that will not be the objective of the sort of community activity now envisaged and being supported by the government. The implications of such a development are far more profound than it would at first appear.

There is growing evidence that in sectors of the government as well as in the boardrooms of the massive multinational corporations, there is the intention of actively encouraging the growth of this dual economy. Some of the corporations have already released leading executives to become involved in job-creation schemes. As one of them jokingly put it, 'The government can only pay about

£8000 for them to do this work, so we'll make up the other £30,000 for them.'

Their concern seems to be to diffuse the growing resistance among the unemployed and the critical sectors of the community to the manner in which the corporations and the government are displacing large groups of workers on the one hand and dominating the manner in which technology is developing on the other.

In the process they are, of course, maximising their profits and in practice are able to circumvent or negate any enlightened programmes of industrial reorganisation which nation-state governments might attempt to implement. Such enlightenment, I must add, is not particularly evident anywhere in this country under this government or indeed the last one. It is not, therefore, too farfetched to suggest that the government and the large employers will conspire to force the unemployed and the deprived communities to provide their own social services.

Culturally, the members of these communities may live in a world they no longer understand or can cope with. At a political level, with such a concentration of economic power and technical know-how within the elite, it is unlikely that the present concepts of equality and democracy would long survive, and the development of a highly centralised, authoritarian corporate state would thereby be facilitated. Further, the small elite in the highly capitalised 'scientific and productive' sector of the economy would be involved in the design and development of forms of repressive technology which could be used against the remainder without the hindrance of any countervailing force in the traditional form of a class-conscious, organised working class within the productive processes.

A FOOT IN BOTH CAMPS

It is against this scenario that the Lucas workers have raised the demand that everybody should have the right to a job, and on socially useful work. If there is to be a dual economy with high-level and low-level sectors, then the work in the advanced sector should be available to, and shared out among, the whole labour force. No one who wants to take part should be written off as

Some Social and Technological Projections

incompetent or incapable, and an appropriate educational and training infrastructure should be made available to all. Socially useful production should raise the level of interest, involvement and job satisfaction of workers and help to release the immense creativity of the work force which is at present deliberately stifled through Taylorism and scientific management. Work sharing would entail a dramatic reduction of work hours in the 'productive sector' leaving workers time to engage in the alternative 'uneconomic service sector'.

This is not as far-fetched as it may sound. Probably the majority of the population is already engaged in some form of do-it-yourself or voluntary activity not carried on for profit, whether it be community action, hospital visiting, house repairing or home brewing, and there is great scope for cooperative organisation in any such activities where group working is desirable.

All this implies a significant shift in the way in which trade unions would function. Firstly, it would require them to attach far greater importance to workplace organisation than is now the case. The present grudging acceptance of developments like those at Lucas Aerospace would have to be replaced by active support. Secondly, if we are to have a dual economy, the unions must be prepared to function in both sectors, to organise the unemployed and the partially employed and to expand the cooperative sector where traditionally 100 per cent trade unionism has been readily accepted. To succeed in doing this, a dramatic re-examination of trade-union structures will be necessary, and the highly centralised authoritarian bureaucratic forms of some of them will have to be altered. More importantly, they could provide a bridge through both education and collective bargaining whereby workers could as of right have a foot in both camps if we are going to accept that there will be two. In other words, they would be helping to link together the dual role of human beings, namely that of producers and consumers.

ON WORKERS' CONTROL

If we look about us in Britain at the moment, we can see there is a serious crisis. There is a crisis of structural unemployment, our

Architect or Bee?

National Health Service is being dismantled and even the air we breathe is being polluted by this system. Yet, in spite of this crisis, the influence of the Left is significantly small in Britain.

When you say that to left-wing people, they have a tendency, in all seriousness, to take the view of Brecht who, on one famous occasion, said, with brilliant irony, 'The government has decided the people are wrong, therefore the people must be disbanded.' Some people say if *only* we had the French working class or if *only* we had the Italian working class we would surge forward. But we have the British working class with all its weaknesses and all its strengths – and its strengths are many. It is an experienced and courageous working class.

Part of the trouble is that we don't listen to the working class nearly enough. When you talk to workers about a socialist society, they tend to ask whether there is any country where that sort of society exists, and that's a pretty difficult question. I wouldn't like to say that the sort of society I want is the one I see in the Soviet Union, for example. It is in many respects the contrary of what I wish to see.

They ask 'What kind of leaders? How would the country be run?' It is clear from the whole way they question these things that they have no intention of replacing one elite with another. They don't want tsars, whether they be government tsars, trade-union tsars or any other kind. They want a society in which they can really participate and use their creativity to the full.

Now in many ways, that concept is in contradiction with the notion of leadership that exists in many parties and factions in the United Kingdom. There is the notion going around that leadership is declaring yourself to be a vanguard. Having said that, you then pursue a sort of Jesuitical logic and say that if we are the vanguard it necessarily follows that we represent the highest level of consciousness of the working class. We are also the most dynamic in the working class and anybody who disagrees with us must be an enemy. This dogma induces political leaders to perceive their role as solely that of telling others what to do, since by virtue of being part of this elite vanguard, they know by definition precisely how ordinary people should behave in all circumstances and at all times.

The problem about this is that even if we could find leadership – and I certainly don't see it in Britain at the moment – such a leadership would deny the working class a most precious experience. This is the self-activism and development which raise the level of consciousness and competence to that high level which is a prerequisite of social advance, that is, a really democratic society. Insofar as workers' control is one of the components which would do that, it is, in my view, important.

WORKERS HAVE VISION

I mention that particularly because of the significant developments in Lucas Aerospace. As a result of trying to draw an image of what the future of that company might be like, and an image of the society in which it would operate, it has been possible to be practically engaged in a whole series of activities which have exposed our members, far more clearly than ever before, to the objective role of the government and many of the ministers within it. We have seen both the pathetic, slavish grovelling of some government ministers to this big multinational company in which they work, and the objective role of the trade-union leaders who, behind our backs, had secret meetings with the company to prepare the carve-up of our jobs. These are the same people who, when confronted eight years ago with our request to assist us in deciding what products we could be making, and how we should be making them, were absolutely silent. Yet they are the people who will tell us that they know best and will always lead us. We don't like that kind of hierarchy at all, whether it be in the trade-union movement or in political parties.

In the course of that struggle, we have, as well, been able to demonstrate in practice the non-neutrality of science and technology. We have been able to do it in objective circumstances through activity, in a way we could never have done just by reading about it, or by getting lectures from very profound leaders. 'I know because I do,' as one of the Lucas workers said, 'not, I do because I know.'

Any organisation that provides a framework in which workers can be involved in that kind of activity is, in my view, an important and significant development towards creating the level of consciousness we need to safeguard and guarantee democracy in the future.

STAY AT THE BASE

When the superstructure has been changed in other countries, there has been a tendency to run industry precisely as before. It was Lenin who said how important Taylorism could be to the running of the Soviet Union. Now you may recall that Taylor said that a worker should not make improvements upon the instructions given to him, and this reminds me of some political leaders who claim that the mass has been subverted when it does not pursue the direct party line.

What we are talking about is a level of consciousness which comes through struggle. We maintain that those who become separated from the base of the action and go into the superstructure very quickly begin to challenge the base. There is a contradiction between superstructure and base. In my view, any full-time trade-union leader, taken away from the point of production – he can be as benevolent and dynamic and energetic and political as he wants – that person, over five years or so, will change. I have seen that change happen among my colleagues. If we are talking therefore about workers' control, we have got to ensure that we develop mechanisms where people are continuously exposed to the contradiction at the point of production itself.

When Jack Mundy suggested that trade-union leaders should return to the point of production after three years and work there for six years before they have the possibility of becoming a full-time official again, he was torn apart both by the Right and the Left in Australia. Both these groups saw real workers' control, where there was rotation of function and involvement of people at the point where the decisions would be made, as a challenge to their power structures. It is worth recalling that during the Cultural Revolution in China (which, in spite of some horrific excesses, raised profound questions of the contradictions between base and superstructure) the workers at the Shanghai machine-tool factory said on one occasion that in their opinion, the most dynamic members of a party or class should never go into the superstructure, but should stay at the base, fermenting and toppling the superstructure if necessary. That to me is the dimension of real industrial democracy and workers' control.

COMPROMISE

It may be that industrial democracy and workers' control, as it is spoken about, could represent a compromise with the system which oppresses us. There seems to be in some circles the idea that if only we could get more and more people into positions of authority, we might wake up one morning to find that we have five seats on the board against the employers' four seats and we could then disband it altogether.

I don't believe for one moment that any ruling class, of the right or the left, acquiesces in its own destruction. I take the view that there is a need ultimately for a party and for an organisation of the working class which can face up to that power. Although that would require a level of consciousness among the working class which we don't have in Britain at the moment, our experience in Lucas Aerospace is that we are developing the levels of consciousness which will make that sort of thing possible.

I think, therefore, that insofar as workers' control can begin to move towards a dual power situation in industry, where the workers begin to flex their own muscles and be conscious of their own great intellects, it is of some importance. These are the people who design and build everything we see about us and without whom we could not survive. After all, you can't live in a pound note, you can't eat a pound note, neither can you drive around in one. All that we see about us comes as a result of the power, the ingenuity and the creativity of working people. If through workers' control they have the opportunity of sensing that power, of using it in practice and thereby understanding how parasitic and how irrelevant are those who control society, then workers' control will have been important. Insofar as it represents one challenge to the naked power of the multinationals in this country, it is, I believe, of great significance.

Architect or Bee?

NUCLEAR POWER – THE POLITICAL DANGERS

It is not my intention to attempt to deal with the issues of nuclear hazards or the wider ecological issues. I will confine myself to what I regard as the enormous political consequences of this type of technology. These political issues can and should represent a major rallying point for the trade-union movement. Nuclear technology will be an avenue to attack basic democratic and trade-union rights, which our movement has established after generations of struggle and sacrifice.

For 300 years, our predecessors fought against the employers, governments and the law to establish the right to strike. Only a slave cannot strike! Nuclear technology will prove to be one of the most effective strikebreakers in history. It will be recalled that when previous governments, both Conservative and Labour, sought to deny some of these rights through industrial-relations policy, there was an upsurge against it. With this type of technology, the same sort of thing will be done by more surreptitious means. We will therefore be able to find a community of interest between those concerned about the ecological impacts and hazards of nuclear power, and those concerned about its impact on democratic rights. So far, progressive forces have failed to make these links, and one witnesses on all sides narrow, single-issue activity, lacking a cohesive organisational and political framework.

These concerns and the manner of addressing them do not fall into conventional left or right political pigeonholes. We find Margaret Thatcher's approval of the French socialist government's nuclear programme, and the adherence of the Soviet Union to its nuclear policies in spite of the Chernobyl accident. This compares strangely with the situation in the United States, where the building of nuclear reactors stopped after the Three Mile Island events. All these countries of course continue to develop nuclear power for their military programmes.

The nature of this technology will shape the society in which we live, and even if we could succeed in so redesigning the nature of that industry as to render it 'safe' in an environmental sense, it is unlikely that we will be able to render it 'safe' from a democracy point of view. Tony Benn, an advocate of industrial democracy, was the minister for energy and felt compelled to use troops to end a dispute in the nuclear industry on the grounds that the strike constituted a major public hazard. It is now significant that arising from the public concern about nuclear power which has been heightened dramatically by the Chernobyl accident, Tony Benn and some other Labour leaders have been honest enough to alter their position on the nuclear-power issue and have now pressed for a searching and public debate on the phasing out of nuclear power plants in Britain.

Once it has been established through the nuclear industry that it is industrial policy to prevent strikes in circumstances of this kind, the argument could then extend to many ICI plants, or indeed to almost any large plant, as we know from the recent events in the north of Italy. This kind of industry could do to us what anti-trade-union legislation has failed to do to us in the past, that is, deny us a most basic right – the right to strike.

BRITISH *BERUFSVERBOT*

Trade unions have always tried to prevent employers victimising workers because of their political views. With nuclear technology, it will be said that to guard against terrorists, the government must increasingly insist on knowing the political affiliations, the intimate personal habits and even the bank balances of those who work in those industries. It will be a massive intrusion into the personal privacies of those who are expected to do this work, and workers will be denied the right to work in such industries on the basis of the political views that they hold. We will no longer be able to sneer at West Germany, with its repressive legislation. We will have our own, much more subtle, English forms of *Berufsverbot*.

Even the communities which live around the stations will be subjected to scrutiny, in case they might harbour potential terrorists.

For those who work in the industry, there can be no industrial democracy. Since the industry and its operation represent enormous industrial hazards, all actions of the workers in it are determined well in advance. All command systems emanate centrally, and must be obeyed at all levels. It is run on almost military lines. Even the clothing that the workers wear is specified in many of the areas.

One of the latest arguments is to say that we must have this technology because it will create new jobs for us. We suddenly find lots of strange allies of the working class; those converted overnight to concern about our right to work. We heard little from these people when Arnold Weinstock was destroying 60,000 jobs in GEC. We heard little from them when companies like Leyland were destroying thousands of jobs as well. In particular, we in Lucas Aerospace heard nothing from them when our company brutally reduced its work force from 18,000 to 12,000. If for a moment we accept that these people are genuinely concerned about jobs for the whole population, we would have to say immediately that their proposal that we need nuclear power to do this will represent the most expensive job-creation scheme in history. Probably in more senses than one.

If we regard this expense, not in terms of hazards or potential loss of life, but simply in first-order economic terms, it will cost something like £600,000 to create one permanent job at Sellafield. Yet one could create jobs in energy conservation, say in the East End of London, insulating houses at approximately £4000 per job.

LEARN TO USE THE ENERGY WE HAVE!

I wish to state, as a trade unionist and a technologist, that I am not opposed to technological change. I am certainly not like some romantics who seem to believe that before the Industrial Revolution, the populace spent its time dancing round maypoles in unspoiled meadows. I am deeply conscious of the enormous contribution science and technology have made towards eliminating squalor, disease and filth. What I am totally opposed to is the irresponsible use of technology, and I regard it as irresponsible to introduce a form of technology such as nuclear power, until we have examined very carefully what real alternatives exist – until we have put as much research money into alternative forms as we presently put into nuclear power. Further, we must test and assess the long-term implications of some of the nuclear technologies at present being proposed.

Some Social and Technological Projections

Nuclear power, as we now know it, will not create the type of jobs we should be demanding in the trade-union movement. They will be hazardous jobs, hazardous for the workers and for the community. There will be enormous political implications in the social infrastructure which will be set up around these industries. It *will* destroy our right to strike.

Those who work at the rank-and-file level in the trade-union movement have an enormous task to get these issues raised in that movement, and gradually to get opposition built up to it.

ORDINARY PEOPLE

I am frequently asked if I believe that ordinary people are really able to cope with the complexities of advanced technology and modern industrial society.

I have never met an ordinary person. All the people I meet are extraordinary. They are fitters, turners, housewives, nurses, airline pilots, doctors, draughtsmen, designers, teachers. They all bring vast bands of intelligence, experience and knowledge to the daily tasks that they perform.

Lucas Aerospace has ordinary maintenance fitters who go to London airport if a generator system is causing a problem. The whole aircraft might perhaps be grounded because of it. One of these fitters can listen to the generator, make a series of apparently simple tests – some of the older fitters will touch it in the way a doctor will touch a patient – and, if it is running, will be able to tell you from the vibrations whether a bearing is worn and which one.

The fitters will subsequently make decisions about the reliability of that piece of equipment, and on that decision, the lives of 400 people may directly depend. The decision may be far more profound in many ways than that which a medical practitioner might make, yet if you asked those 'ordinary people' to describe how they reached that decision, they could not do so in the usually accepted academic sense. That is to say, they would not be capable of drawing a decision-making tree leading to their final conclusion. Yet that conclusion will be right, because they have spent a lifetime accumulating the skill and knowledge and ability which helps them to arrive at it.

Architect or Bee?

When a great politician goes on a world tour, he or she will be dependent, among other things, on the skill and ability of the people who have maintained the aircraft, those who designed and built it, the pilots who fly it and the traffic controllers who regulate its flight paths. However, all these will be completely unheard-of, ordinary people.

Likewise, when we travel on a high-speed train, we are dependent on the skill and ability of those who have maintained and built it, the people who maintain the tracks and the people who operate the signalling systems.

In everything we do, whether we are in hospital, travelling along a motorway or in the underground, we depend on the skill, ability, understanding and intelligence of so-called ordinary people. Every building we see about us has been constructed by people like that. Every car that runs along the roads has been built by them. Yet, in spite of all the knowledge these people demonstrate in practice, they are effectively excluded from the major decisions which affect the way their lives are run and the industries in which they work are organised. They are induced to believe that they are incapable of making the major decisions about the way society should develop.

This in spite of the fact that everything we see about us has been designed and built by those people. They are deliberately conditioned to feel no association between the technology, or the products they have produced, and their own intelligence.

INTELLIGENCE OR LINGUISTIC ABILITY?

When building workers erect a building, they do not scratch their name on it as an artist or a sculptor might do. They do not associate themselves with the building, yet it has been produced by them. The whole educational and political system works to reinforce these assumptions. We have more regard for those who write and talk about things than those who actually do them. We confuse linguistic ability with intelligence.

Workers express their intelligence by the things they make and do, and the manner in which they organise themselves in producing these things. They have developed and utilised very high-level and complex communication systems among themselves.

Some Social and Technological Projections

When, for example, they are erecting a power station and are lining up a turbine and generator set, the workers will go through highly complex decision-making routines, and communicate these decisions to each other with crisp, simple sentences. An instruction manual on how this is done could be a vast technical work. If you hear two 'intellectual workers' talking about these procedures, they frequently have to go into great detail and describe these matters in a language of great complexity.

It seems to me, therefore, that if we are effectively to question the way science and technology are developing, and do it in such a manner that we involve masses of people in the process, avoiding the dangers of elitism (which are as dangerous from the Left as from the Right), we shall have to organise our affairs so as to release the tacit knowledge of these workers.

Further, we shall have to organise our decision-making routines and our social organisations in such a way that the knowledge, intelligence and experience of these people is not bludgeoned into silence by academic rambling and technological jargon and a deliberately confusing overcomplexity.

This is not to say that profound questions can be treated in a simplistic fashion, but they should be dealt with in a manner which makes them accessible to so-called ordinary people. These questions are of profound and fundamental importance to the whole issue of democracy itself. If we hold it to be desirable that our societies are composed of alert, vigilant, self-active, self-reliant, cooperative and concerned citizens, we have to provide political structures which cater for this.

It is my experience, having spent some twenty-five years in the engineering and manufacturing industries, that 'ordinary people' are well capable of understanding and coping with these problems when they are directly put to them. If academics have difficulty in communicating with workers, then it is *their* fault, not the fault of the workers involved. As an African rebel leader once said, 'Let your words be so direct and clear and simple that the ideas they represent flow through ordinary people's consciousness as naturally and as easily as the wind and the rain flow through the woods.'

Architect or Bee?

COULD WE USE SCIENCE DIFFERENTLY?

I have suggested earlier that science and technology are not neutral, but reflect the economic base which gave rise to them. If this is correct, then the use/abuse model will be inadequate to explain the contradictions we see in technologically advanced society. We shall have to probe deeper.

Technological change has certainly been used, not merely to increase productivity, but to extend control over those who work within those processes. Further, D. F. Noble has brilliantly demonstrated that engineers, in the application of science and technology, have been serving and advancing the cause of corporate capitalism.[1]

I have questioned whether the means of production so developed would be appropriate in a society where human beings could develop their potential to the full, even when the ownership of the means of production is 'in the hands of the people'.

Science as practised in the technologically 'advanced' nations, and I would include here the so-called socialist countries, shares with Taylorism the methodological assumptions of predictability, repeatability and quantifiability.

If one accepts these to be tenets of the scientific method, it then follows that to be scientific implies eliminating human judgement, the subjective and uncertainty; yet skill, in the intellectual and also in the manual sense, can be closely related to the ability to handle uncertainty. Skilled work, we may say, is work of risk and uncertainty, whereas unskilled work is work of certainty. The contrast between a skilled turner using a universal lathe and an unskilled worker on a numerically controlled machine will illustrate the point, as will the contrast between a conventional designer and one using a design-manual type of computer-aided design system. Further, the exercise of skill is an important learning and developing process. If we regard it as desirable to enhance human skill and ability, we have to design systems which are responsive to human judgements, and which respond to the persons using them rather than acting upon them. The telechiric devices described earlier begin to address this problem. Other ideas are being

Some Social and Technological Projections

explored which in embryo start to point the way to a human-enhancing, liberatory form of technology. Two examples only will be given here to illustrate the possibilities – one in the field of manual work, the other in the field of intellectual work.

FIRST EXAMPLE

Over the past 200 years, turning has been one of the highly skilled jobs to be found in most engineering workshops. Toolroom turning is one of the most highly skilled jobs of all. The historical tendency, certainly since the war, has been to deskill this function by using numerically controlled machines. This is done by part-programming, a process by which the desired numerically controlled tool motions, are converted to finished tapes. Conventional (symbolic) part-programming languages require that a part-programmer, having decided how a part is to be machined, describes the desired tool motions by a series of symbolic commands. These commands are used to define geometric entities, that is, points, lines and surfaces which may be given symbolic names.

In practice, the part-programming languages require the operator to synthesise the desired tool motion from a restricted available vocabulary of symbolic commands. However, all this is doing is attempting to build into the machine the intelligence that would have been exercised by a skilled worker in going through the labour process.

It is possible, by using computerised equipment in a symbiotic way, to link it to the skills of a human being and define the tool motions without symbolic description. Such a method is called analogic part-programming.[2] In this type of part-programming, tool-motion information is conveyed in analog form by turning a crank or moving a joystick, or some other hand/eye coordination task, using readout with precision adequate for the machining process. Using a dynamic visual display of the entire working area of the machine tool including the workpiece, the fixturing, the cutting tool and its position, the skilled craftsman can directly input the desired tool motions to 'machine' the workpiece in the display.

Such a system, which may be described as part-programming by doing, would represent a sharp contrast to the main historical tendency towards symbolic part-programming. It would require no knowledge of conventional part-programming languages, because the necessity to describe symbolically the desired tool motions would be eliminated. This is achieved by providing a system whereby the information regarding a cut is conveyed in a manner closely resembling the conceptual process of the skilled machinist. Thus it would be necessary to maintain and enhance the skill and ability of a range of craft workers who would work in parallel with the system.

Significant research has been carried out in this field,[3] yet, in spite of its obvious advantages, it has not been received with any enthusiasm by large corporations, or indeed funding bodies. This would appear to be an entirely political judgement rather than a technological one.

SECOND EXAMPLE

In the field of intellectual work, Rosenbrock has questioned the underlying assumptions of the manner in which we are developing computer-aided design systems. He charges that the present techniques fail to exploit the opportunity which interactive computing can offer. The computer and the human mind have quite different but complementary abilities. The computer excels in analysis and numerical computation. The human mind excels in pattern recognition, the assessment of complicated situations and the intuitive leap to new solutions. If these different abilities can be combined, they amount to something much more powerful and effective than anything we have had before.

Rosenbrock objects to the automated manual type of system, since it represents, as he says, 'a loss of nerve, a loss of belief in human abilities, and a further unthinking application of the doctrine of the "Division of Labour".'[4]

As in the case of turning, described above, Rosenbrock sees two paths open in design. The first is to accept the skill and knowledge of the designer, and to attempt to give designers improved techniques and improved facilities for exercising their knowledge and skill. Such a system would demand a truly interactive use of

Some Social and Technological Projections

computers in a way that allows the very different capabilities of the computer and the human mind to be used to the full.

The alternative to this, he suggests, is 'to subdivide and codify the design process, incorporating the knowledge of the existing designers so that it is reduced to a sequence of simple choices'.[5] This, he points out, would lead to a deskilling, so that the job could be done by a person with less training and less experience.

Rosenbrock has demonstrated the first human-enhancing alternative by developing a CAD system with graphic output to develop displays from which the designer can assess stability, speed of response, sensitivity to disturbance and other properties of the system. (See Figure 17.)

If, having looked at the displays, the user is not satisfied with the

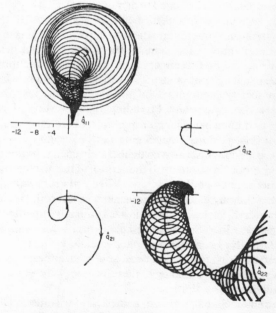

Fig. 17. Graphic display of mathematical functions.

performance of the system, the displays will suggest how it may be improved. In this respect the displays carry on the long tradition of early pencil and paper methods, but of course they bring with them much greater computing power. Thus, as with the lathe and the skilled turner, so also with the VDU and the designer: possibilities exist of a symbiotic relationship between the worker and the equipment. In both cases, tacit knowledge and experience are accepted as valid and are enhanced and developed.

In Rosenbrock's case, it was necessary to examine the underlying mathematical techniques involved in control-systems design.[6] The outcome of his work does demonstrate in embryo that there are other alternatives if we are prepared to explore them before we close off options – the 'Lushai Hills effect'.[7]

HUMAN-ENHANCING

These examples have been cited to demonstrate that it is possible so to design systems as to enhance human beings rather than to diminish them and subordinate them to the machine. It is my view that the development of systems of this kind, however desirable they may be, will be fiercely opposed and vigorously suppressed since they challenge power structures in society. Those who have power in society, epitomised by the vast multinational corporations, are concerned with extending their power and gaining control over human beings rather than with liberating them.

It is not suggested here that engineers who design conventional systems are hideous fascists who deliberately engage in this design in order to subordinate others to the control of the machine and the organisations that own the machine. What I am suggesting, however, is that they are dangerously mistaken in regarding their work as being neutral. Such a naive view was ruthlessly exploited in the Third Reich as Albert Speer pointed out in his book *Inside the Third Reich*: 'Basically, I exploited the phenomenon of the technician's often blind devotion to his task. Because of what appeared to be the moral neutrality of technology, these people were without any scruples about their activities.'

Science and technology are not neutral, and we must at all times expose their underlying assumptions. We can, at the same time,

Some Social and Technological Projections

begin to indicate how science and technology might be applied in the interests of the people as a whole, rather than to maximise profits for the few.

But it is not only the scientist and entrepreneur who must be more responsible. One of the saddest things for me has been the inability of the trade-union movement to foresee the impact that new technology would have on the unions as institutions, and particularly on their members. Even more distressing was the unwillingness even to discuss the issues in any meaningful way; and finally, when the Lucas workers sought to do something, it was deplorable to witness the manner in which the trade-union bureaucracy (with a few honourable exceptions) converged to undermine what they perceived to be an 'oppositional movement'. Furthermore, they lost a unique opportunity to show that the trade-union movement could modify and transform itself, in the terms of Wainwright and Elliott, into 'a new trade unionism in the making'.[8]

The opportunity was there to demonstrate that they were concerned about environmental issues, that they were concerned about the communities in which their members worked, and not just their members, and that in dealing with some of their own problems of growing structural unemployment they could use their skills and talents to support less fortunate sectors of the community. They lost a tactical advantage which allowed Thatcher subsequently to portray the unions as entirely selfish, inward-looking organisations lacking in compassion and concern.

However, even at the level of narrow self-interest, they failed to grasp the opportunity to organise their affairs and mobilise their strength to be one step ahead of the multinationals in defining the types of human-centred technologies that they required, rather than slavishly reacting to steps the employers had already taken. The basis existed in the Lucas Workers' Plan for avoiding, or at least preparing for, a Wapping-like situation. That possibility still exists for many engineering workers and those in a range of white-collar and administrative areas, but time is not on their side.

Trade unionists, socialists, liberals and humanitarians are now beginning to pay the price for ignoring technological development.

Some have been obsessed with the contradictions of distribution to the exclusion of an ongoing concern about the contradictions of production. Others are not interested in 'industry' except in a voyeurist fashion and many seem to believe that the world can be composed entirely of well-meaning sociologists.

A deep analysis of the form and nature of technology is required, together with a recognition that whether we like it or not it does at this stage constitute a leading edge in society in rather the same way religion did at earlier stages.

There is evidence that the Lucas Workers' Plan has contributed – at least to a meaningful discussion – on some of these issues, and in trade-union terms, it does represent quite a shift forward. The International Metalworkers' Federation, which represents over 150 unions in some seventy countries, has produced a report in seven languages which it is distributing free of charge in order that these discussions may take place among its membership.[9]

There is no doubt that at the political level, the developments in the GLC in setting up the Greater London Enterprise Board were a further expression of the ideas formulated at Lucas. The Lucas Workers' term 'socially useful production' is now in the life stream of large sections of the political movement.[10] The technology networks, for all their unevenness and considerable difficulties, represent an important step forward for that concern in the Lucas Plan, the democratisation of science and technology, and the involvement of large sections of the community in defining the types of products and services they require.[11] Although these activities were to take place at a local and even individual factory level, they were to be part of a wider scheme as set out in the London Industrial Strategy,[12] and based on the skills of the people of London as set out in the Labour Plan.[13]

These were but the job-creation and socially useful production aspects of the plan permeating the political body. In parallel, there was a paradigmatic shift in our concept of how systems might be designed in a human-centred way. An outgrowth of the earlier Rosenbrock work described above was the two-year programme to design and develop a human-centred lathe in which the qualitative, subjective elements would be handled by the operator and the quantitative elements by the machine. Such a control system and its interface were developed at UMIST.[14]

This in turn laid part of the basis for the GLEB-sponsored ESPRIT project to design and build the world's first human-centred computer-integrated manufacturing system.[15] Discussion with colleagues in other countries since the late 1970s has gradually built up a network of engineers, scientists, philosophers, researchers and social scientists who are now working on these ideas from a theoretical level through to very practical projects. Some of them, like Peter Brödner, are discussing how we might conceptualise the factory of the future – an anthropomorphic factory – described in his vitally important book which I hope will soon be available in the English language.[16] Given the deluge of quasi-expert systems, artificial-intelligence software tools and frameworks for fifth-generation systems, it is important to have a forum where these matters can be discussed at both a theoretical and practical level, and Rajit Gill and others (including the author) have founded the journal *AI and Society – Human Centred Systems Journal*.[17] Likewise, international bodies dealing with computerisation are setting up social effects committees to analyse the multiplier effects of technology, and the International Federation of Automatic Control is a good example of this.

These all constitute important developments which I hope will converge adequately to produce visible and meaningful alternatives before an infrastructure of machine-based systems, with its attendant reconditioning through 'training', permanently closes off options for human-centred technology in rather the same way as a similar option in respect of skilled manual work was closed off around the fifteenth and sixteenth centuries.

The rate of technological change suggests to me that we have a mere fifteen or twenty years in which to do so, otherwise we will find, as an artificial-intelligence expert put it recently on British television, that humans will have found their natural place in the evolutionary hierarchy. Namely animals at the bottom, human beings in the middle and thinking machines at the top!

The future is not 'out there' in the sense that a coastline is out there before somebody goes to discover it. The future doesn't have predetermined shapes and forms and contours. The future has yet to be built by people like you and me and we do have real choices. I hope the ideas contained in this book will have highlighted at least some of them.

The choices are essentially political and ideological rather than technological. As we design technological systems, we are in fact designing sets of social relationships, and as we question those social relationships and attempt to design systems differently, we are then beginning to challenge, in a fundamental way, power structures in society.

The alternatives are stark. Either we will have a future in which human beings are reduced to a sort of beelike behaviour, reacting to the systems and equipment specified for them, or we will have a future in which masses of people, conscious of their skills and abilities in both a political and technical sense, decide that they are going to be the architects of a new form of technological development which will enhance human creativity and mean more freedom of choice and expression rather than less. The truth is, we shall have to make the profound political decision as to whether we intend to act as architects or behave like bees.

REFERENCES

CHAPTER 1 *Identifying the Problem*

1 Cooley M. J. E. 'The Knowledge Worker in the 1980s', Doc. EC35, Diebold Research Programme, Amsterdam, 1975.
2 Braverman H. *Labor and Monopoly Capital. The Degradation of Work in the 20th Century*, Monthly Review Press, New York, 1974.
3 Dreyfus and Dreyfus, *Mind over Machine*, Glasgow, 1986.
4 Bodington S. *Science and Social Action*, Allison & Busby, London, 1979.
5 Needham J. 'History and Human Values' in H. and S. Rose (eds), *The Radicalisation of Science*, Macmillan, London, 1976.
6 Cooley M. J. E. 'Computer Aided Design, Its Nature and Implications', AUEW-TASS, 1972.
7 Polanyi M. 'Tacit Knowing: its bearing on some problems of philosophy', *Review of Modern Physics*, Vol. 34, October 1962, pp. 601–605.
8 Maver T. W. *Democracy in Design Decision Making CAD*, IPC Science and Technology Press, Guildford, Surrey, 1972.

CHAPTER 2 *The Changing Nature of Work*

1 *Economist*, 22 January 1972
2 *Daily Mirror*, 7 June 1973.

CHAPTER 3 *The Human–Machine Interaction*

1 Cooley M. J. E. 'Criteria for Human Centred Systems' in *A.I. and Society*, London, 1987.
2 PROC 'Human Choice and Computers', Report HCC, Lp.5, IFIP, Vienna, 1974.
3 Kling R. 'Towards a People Centred Computer Technology', Proc. Assoc. Computer Mach. Nat. Conf., 1973.
4 Boguslaw R. *The New Utopians: A Study of Systems Design and Social Change*, Prentice-Hall, New Jersey, 1965.
5 Taylor F. W. *On the Art of Cutting Metals*, 3rd edition revised. ASME, New York, 1906.
6 *Dataweek*, 29 January 1975.
7 'Nissan Agrees with Unions on Robots', *Computing*, 10 March 1983, p. 9.
8 Fairbairn W. quoted by J. B. Jefferys, *The Story of the Engineers*, Lawrence & Wishart for the AEU, 1945, p. 9.
9 *Engineer*, 20 June 1974.

10 *Economist*, 14 July 1973.
11 Shakel B. 'The Ergonomics of the Man/Computer Interface', Proc. Conf. Man/Computer Communication, Infotech International Ltd, Maidenhead, UK, November 1978, p. 17.
12 Faux R. *The Times*, 26 March 1975.
13 Rose S. *The Conscious Brain*, Penguin Books, 1976.
14 Archer L. B. *Computer Design Theory and the Handling of the Qualitative*, Royal College of Art, London, 1973.
15 Nadler G. 'An Investigation of Design Methodology Management', *Science* Vol. 3, June 1967, pp. 642–655.
16 Lobell J. 'Design and the Powerful Logics of the Mind's Deep Structures', DMG/DRSJ, Vol. 9, No. 2, pp. 122–129.
17 Beveridge W. I. B. *The Art of Scientific Investigation*, Mercury Books, London, 1961.
Eisley L. *The Mind as Nature*, Harper & Row, New York, 1962.
Fabun D. 'You and Creativity', *Kaiser Aluminum News*, Vol. 25, No. 3.
18 Marx K. *Capital*, Vol. 1, p. 174, Lawrence & Wishart, London, 1974.
19 Silver R. S. 'The Misuse of Science', *New Scientist*, Vol. 166, p. 956, 1975.
20 Rose S. 'Can Science Be Neutral?', Proc. Royal Institute, Vol. 45, London, 1973.
21 Rose H. & S. 'The Incorporation of Science', in H. and S. Rose (eds), *The Political Economy of Science*, Macmillan, London, 1976.

CHAPTER 4 *Competence, Skill and 'Training'*

1 Braverman op. cit.
2 Dreyfus & Dreyfus op. cit.
3 Kantor *Vorlesungen über Geschichte der Mathematik*, Vol. 2, Leipzig, 1880.
4 Olschki *Geschichte der neusprachlichen Wissenschaftlichen Litteratur*, Leipzig, 1919.
5 Sohn Rethel A. *Intellectual and Manual Labour: A Critique of Epistemology*, Macmillan, London, 1978.
6 Bowie T. *The Sketchbook of Villard de Honnecourt*, Indiana University Press, 1959.
7 Kemp M. *Leonardo da Vinci – The Marvellous Works of Nature and Man*, J. M. Dent & Sons Ltd, London, 1981, p. 26.
8 Ibid.
9 Polanyi op. cit.
10 Kemp op. cit., p. 102.
11 Cooley M. J. E. 'Some Social Implications of CAD' in Mermet (ed.), *CAD in Medium-Sized and Small Industries*, Proceedings of *MICAD 1980*, Paris, 1980.
12 Cooley M. J. E. 'Computerisation – Taylor's Latest Disguise' in *Economic and Industrial Democracy*, Vol. 1, Sage, London and Beverly Hills, 1981.
13 Weizenbaum J. *Computer Power and Human Reason*, W. H. Freeman & Co., San Francisco, 1976.

References

14 Aspinal, Cooley et al. *New Technology, Employment and Skill*, Council for Science and Society, London, 1981.
15 Rosenbrock H. H. *Computer Aided Control Systems Design*, Academic Press, London, 1974.
16 Cooley M. J. E. 'Trade Unions, Technology and Human Needs', a 50-page report available free in seven languages from the International Metalworkers' Federation.
17 'Human Centred Robot', *Financial Times*, 4 February 1986, p. 10.
18 *Shooting Life*, Spring 1987, p. 11.
19 Taylor F. W. op. cit.

CHAPTER 5 *The Potential and the Reality*

1 'Shiftworking and Overtime Practices in Computing', Rep Computer Economics Ltd, Richmond, Surrey, 1974.
2 Mott P. E. *Shiftwork; the Social, Psychological and Physical Consequences*, Ann Arbor, 1975.
3 Rosenbrock H. H. 'The Future of Control', *Automatica*, Vol. 13, 1977.
4 Östberg O. 'Review of Visual Strain with special reference to microimage reading', International Micrographics Congress, Stockholm, September 1976.
5 Allen B. 'Health Risks of Working with VDUs', *Computer Weekly*, 9 February 1968, p. 3.
6 Report, *New York Times* Survey NIOSH, New York, 1976.
7 Östberg O. 'Office Computerisation in Sweden. Worker Participation workplace design considerations and the reduction of visual strain', Proc. NATO Advanced Studies Institute on Man, *Computer Interaction*, Athens, September 1976.
8 'Making Sure Technology Is Right for the Press', *Computing*, 23 March 1978, p. 74.
9 'Electronic Office System Designed to Improve Managers' Productivity', *Computer Weekly*, 21 December 1978, p. 12.
10 Act relating to Worker Protection and Working Environment, Order No. 330, Statens Arbeidstilsyn Direktoratet, Oslo.
11 Urquart A. *Familiar Words*, cited in Marx K. *Capital*, London, 1855; Lawrence & Wishart, London, 1961, Vol. I, p. 36e.
12 Smith A. *The Wealth of Nations*, Random House, New York, 1937.
13 Martyn H. *Consideration upon the East India Trade*, London, 1801.
14 Braverman H. op. cit.
15 Dochery P. 'Automation in the Service Industries', Round Table Discussion, IFAC, 1978.
16 Kraft P. *Programs and Managers – The Routinization of Computer Programming in the United States*, Springer Verlag, Berlin, Heidelberg, New York, 1977.
17 Babbage C. *On the Economy of Machinery and Manufactures*, New York (reprint), 1963.
18 Carlson H. C. in Braverman, op. cit.

19 *Academy of Management Journal*, Vol. 17, No. 2, p. 206.
20 *Management Science*, Vol. 19, No. 4, p. 357.
21 *Times Higher Education Supplement*, 14 February 1975, p. 14.
22 *New Scientist*, 22 April 1976, p. 178.
23 *Guardian*, 12 October 1979.
24 Marglin S. 'What Do Bosses Do?' in A. Gorz (ed.), *The Division of Labour*, Harvester Press, Sussex, 1976.
25 Hoos I. 'When the Computer takes over the Office', *Harvard Business Review*, Vol. 38, No. 4, 1960.
26 *Realtime*, Vol. 6, 1973.

CHAPTER 6 *Political Implications of New Technology*

1 Rose H. & S. *The Incorporation of Science*, op. cit.
2 Rose H. & S. in W. Fuller (ed.), *The Social Impact of Modern Biology*, Routledge & Kegan Paul, London, 1971.
3 Yankelovich D. *The Changing Values on the Campus*, Washington Square Press, New York, 1972, p. 171.
4 Silver R. S. op. cit.
5 Henning D. Bericht 74-09, Berlin Technical University, 20 January 1974.
6 Jungk R. *Qualität des Lebens*, EVA, Cologne, 1973.
7 Braverman H. op. cit.
8 Lenin V. I. 'The Immediate Tasks of the Soviet Government' (1918) in *Collected Works*, Vol. 27, Moscow, 1965.
9 Cited in *The Division of Labour*, A. Gorz (ed.), Harvester Press, Sussex, 1976.
10 Whyte W. H. *The Organisation Man*, Penguin Books, Harmondsworth, 1960.
11 Marx K. *Critique of the Gotha Programme* ed. C. P. Dutt, Lawrence & Wishart, London, 1938.

CHAPTER 7 *Drawing up the Corporate Plan at Lucas Aerospace*

1 Fletcher R. 'Guided Transport Systems', North East London Polytechnic, 1978.
2 *Engineer*, 14 September 1978, pp. 24, 25.
3 Marglin S. 'What Do Bosses Do?', op. cit.
4 Braverman H. op. cit.
5 Clegg A. 'Craftsmen and the Origin of Science', *Science & Society*, Vol. XLIII, No. 2, Summer 1979, pp. 186–201.
6 Albury D. 'Alternative Plans and Revolutionary Strategy' in *International Socialism*, Vol. 6, Autumn 1979.
7 Nadler G. op. cit.
8 Rosenbrock H. H. 'The Future of Control', *Automatica*, Vol. 13, 1977.
9 Rosenbrock H. H. 'Interactive Computing. A New Opportunity', Control Systems Centre Report No. 338, UMIST, September 1977.
Rosenbrock H. H. 'The Future of Control', op. cit.
10 Weizenbaum J. 'On the Impact of the Computer on Society, How does one

References

insult a machine?' *Science*, Vol. 176, 1972, pp. 609–14.

Weizenbaum J. *Computer Power and Human Reason*, W. H. Freeman & Co., San Francisco, 1976.

11 Cooley, Friberg, Sjöberg *Alternativ Produktion*, Liberförlag, Stockholm, 1978.

CHAPTER 8 *The Lucas Plan – Ten Years On*

1 Booklets and videotapes from GLEB, 63/67 Newington Causeway, London SE1.
2 Cooley M. J. E. and Murray R. Report No. IE 413, Tech. Div. GLEB.
3 'Technology Networks', GLEB, 1986.
4 Shelley T. 'Solid Rubber Tyre Perfected at Last', *Eureka*, Vol. 6, No. 2, February 1986, pp. 34–6.
5 Craven F. 'Human-Centred Turning Cell', RD Projects, London, October 1985.
6 Cooley M. J. E. 'Trade Unions, Technology and Human Needs', op. cit.

CHAPTER 9 *Some Social and Technological Projections*

1 Noble D. F. *America by Design*, Alfred A. Knopf, New York, 1977.
2 Gossard D. & von Turkovich B. 'Analogic Part Programming with Interactive Graphics', Annals of the CIRP, Vol. 27, January 1978.
3 Gossard D. 'Analogic Part Programming with Interactive Graphics', PhD thesis, MIT, February 1975.
4 Rosenbrock H. H. The Future of Control, op. cit. (Ch. 5).
5 Rosenbrock H. H. 'Interactive Computing: a New Opportunity', Control Systems Centre Report No. 388, UMIST, 1977.
6 Rosenbrock H. H. *Computer Aided Control System Design*, Academic Press, London, New York, San Francisco, 1974.
7 Rosenbrock H. H. 'The Redirection of Technology', IFAC Symposium: Criteria for selecting appropriate technologies under different cultural, technical and social conditions; Bari, Italy, May 1979.
8 Wainwright H. and Elliott D. *A New Trade Unionism in the Making*, Allison & Busby, London, 1982.
9 Cooley M. J. E. 'Trade Unions, Technology and Human Needs', op. cit.
10 Bodington S. et al. (eds) *The Socially Useful Economy*, Macmillan, 1986.
11 'Technology Networks', op. cit.
12 London Industrial Strategy, op. cit.
13 London Labour Plan, GLC, 1986.
14 Rosenbrock H. H. Reports and articles from the Control Systems Dept, UMIST, 1983–6; book by members of the project steering committee (forthcoming 1987).
15 ESPRIT project reports from Technology Division, GLEB, 1986.
16 Brödner P. *Fabrik 2000*, Wissenschaftszentrum, Berlin, 1986.
17 *A. I. and Society*, London, 1987.

INDEX

Absenteeism 34–35
Abstraction, level of 53, 73
Academic Department, operation of 82–84
Academic titles 61
Accidents 35
Advanced beginners 14
AEI 116
Age factor in recruitment 45–48
Ageing, accelerated by use of VDUs 76
Agriculture 28
Algots Nord, Sweden 137
Alienation 73
Alternative Nobel Prize 4
American Machinist 130
American Society of Mechanical Engineers 80
Analogic part-programming 173
Anthropomorphic factory 179
Anti-science movement 91
Aperture card 17
Appetite loss, due to shift work 75
Apprenticeships 64–68
Aran Islands 34
Architecture, computers in 21, 23, 24
Arms industry, conversion to socially useful production 132
Artificial intelligence 179
  beginning of 55
Artificial limbs, computer design of 21–22
Assembly-line,
  discipline 35
  workers 36
ASTMS 76, 140
Atrophy 43–44
AUEW-TASS 74, 100, 112
Australia 132, 137, 164

Australian Trade Union movement 98
Automatic draughting equipment 17
Automation 2, 9, 58, 158
  myth of 115
Automotive industry 33

Babbage, Charles 79–80
Banking 111
Banks, computers in 32
Barometers 57
Battery-driven car 122–23
Beckett, Samuel 46
Beethoven 52
Benn, Tony 167
Berkeley 112
Berliot 101
Bernholz (design methodologist) 39
Bibliotheque St. Genevieve, Paris 58
Biological effects of VDU use 76
Birmingham 127
Bjorn-Andersen, Niels 87
Boguslaw, Robert 40, 41
Brain, use of 43–44
Braverman, H. 78
Brecht 162
Bremen University 63, 150
Bristol 74
British Leyland 168
British Society for Social Responsibility in Science 104
British Standard Glossary 80
British Telecom 22
Brodner, Peter 179
Brunelleschi, Filippo 59–60
Building design, participation in 23
Burnley 140

California 84

Index

Canada 39
Capital 89
Capital, 134
  as dominant feature in industry 94–95
  organic composition of 28
  short life of 25
Capital-intensive industry 157
Capital-intensive processes 28
Capitalism 1–2, 89, 90, 93–94, 103, 137, 172
Carlson, Howard C. 81
Cell configuration 150
Cell-immobilisation techniques 146
Change, rate of 25–32
Chernobyl 166, 167
China 164
Chrysler 106, 133
Citibank 77
Citroen 101
City University of New York 84
Civil transport 111
Class divisions 108
'Classification and Terminology of Mental Work' 80–81
Clerical jobs, loss of 32
Collective bargaining 161
'College of Business Administration as a Production System' 82
Combine Committee at Lucas 117, 127
Commission for the Future, Australia 132
Communication 22
  speed of 26
Communication protocols 151
Communications systems 115
Community, definition of 130–31, 143
Community armed enterprises 157
Community enterprises 158
Community work 159
Company medicare schemes 159
Competence 70, 82
Competent performers 14–15
Computer-aided design 16, 38, 39, 62, 71–74, 110, 136, 148, 153, 172, 174

Computer-aided manufacturing system 149
Computer-aided production 148, 153
Computer-controlled machines 3
Computer graphics 19
Computer-integrated manufacturing 16
Computerisation 9, 158
  myth of 115
Computer Power and Human Reason 62
Computers, development of 9
Conceptualisation process 23
Concorde 116
Conformity, imposition of 86
Conservation 137
Constipation, due to shift work 75
Consumer, exploitation of 115–16
Consumer skill 63–64
Continuous path milling machine 17
Coons' patch surface definitions 135
Copenhagen School of Economics 87
'Corporation Men' 158–59
County Hall 141
Creativity 38, 48, 52
Cultural Revolution 164
Culture 147
Curricula, nature of 109
Cybernetics 50, 92, 138
Cycleways 146

Danzig 109
Daphne 87
Darwin, Charles 90
Decision-making 14, 49, 169, 171
  democratising of 145
Decision-making process, democratisation of 23
Dedicated machines 109
de Honnecourt, Villard 56–58
Democratisation of decision-making 23
de Montreuil, Pierre 58
Denmark 87, 147, 148, 150
  unemployment in 96
Department of Industry 44, 140
Design,
  as holistic process 51
  changes in 16

Design, history of 54–56
Design,
    quality of 39, 136
    rules for 59–61
Designer 23
Design methodology 38, 50–51, 60, 133
Design process 136
Digitiser 17
Division of labour 3, 80, 112–13, 136, 174
Draughtsman 23
    replacement of 16–17
Dunlop 133
Durer, Albrecht 55–56, 62

Ear protection 21
East Kent Railway line 126
Ecological hazards of nuclear technology 166
Ecologically desirable power unit for cars 122
Ecology, as middle-class preserve 157
Economist 45
Education, deficiencies in 109
Effectiveness 82
Einstein 52–53
Elitism, dangers of 171
Elliott, David 118, 177
EMI 84
Employment, structure of 28
Energy audit systems 145
Energy-conservation 120–22
Energy Exhibition, London 121
Energy Network 143
"Engineer" 45
Engineering, as an art 136
Engraving 65–66
Environment,
    lack of concern for by Trade union movement 177
    work 36
ESPRIT 4, 7, 63, 147–52, 179
Euclidian geometry 56
European Economic Community 32, 63, 96, 116, 140, 147
Exercise 31
Expense accounts 159
Expertise 15

Expo 39
Eye checks for VDU users 77
Eyestrain, due to use of VDUs 76

Fairbairn, Sir William 44, 54, 57
Family life, affected by shift work 75
Fatigue, due to shift work 75
Female characteristics of computerisation 87
Fiat 106, 137
Financial institutions 157
Fixed capital 106–107, 108, 111
    short life of 25
Fletcher, Richard 118, 126, 131
Florence Cathedral 60
Ford Motor Company 74
Fragmentation of skills 73
France 32, 57, 101, 162, 166
    unemployment in 96
Frank-Wolfe algorithm 82
Friedmann, George 101
Fund for the Improvement of Post Secondary Education in Washington 84

General Electric Company 100, 105, 116, 168
General Motors 36, 81, 106, 107
Geometry 56–58
Germany 57, 60, 61
Gill, Rajit 179
Giotto 59
Giovanni, S. 60
Gorz (French political theorist) 36
Graticule 17
Greater London Council 144, 146, 178
    abolition of 142
Greater London Enterprise Board 4, 7, 63, 121, 141–44, 150, 178, 179
Green Bans Movement, Australia 137
Group technologists 36
Gulliver's Travels 49

Harness 48, 62
Health and safety standards 36
Heart valves, computer design of 22
Heath, Professor 41–42

Index

Heat pumps 122, 140
Hegel 16
Heriot-Watt University 41
High-capital equipment 9, 25, 72, 74, 94–95, 100, 102, 110
Hitachi 47
Holistic design 56–59
Holistic similarity recognition 15
Holland 142
Home dialysis machine 127–28
Hounsfield, Godfrey 85
Hours of work 31, 161
Housing 115
Human-Centred Computer-Integrated Manufacturing System 148–56
Human-centred systems, design of 152
Human enhancing 176
Hungary 57
Hunt, Ken 64–66
Hush-kits 146
Hypothermia 114

IBM 85–86, 112
ICI 167
Imagination, importance of 52
Industrial action 111
Industrial democracy 135, 165, 167
Industrial feudalism 157
Industrial militancy 73, 111
Industrial psychologists 36
Industrial relations 42
Industrial reorganisation 160
Industrial Reorganisation Corporation 96, 117
Industrial Revolution 168
Industrial society 114
Industry, Department of 44, 140
Information-retrieval systems 49
Input conditions 19
Inside the Third Reich 176
Insurance 111
Integrated transport system 126–27
Intellectual work 38, 73
Intelligence 42–44, 170–71
Interactive computing 174
Internal combustion engine, ban from city centres 123

International Federation of Automatic Control 43, 179
International Federation of Commercial, Clerical and Technical Employees 77
International Federation of Information Processing 40
International Labour Office 40
International Metalworkers' Federation 178
Investment 141
Ireland, unemployment in 96
Isolation, feeling of by VDU users 77
Italy 35, 95, 106, 137, 158, 162, 167
  unemployment in 96
ITT 105

Japan 35, 42, 47, 95, 124
Jargon 9–10
Jig borer 17
Job-creation schemes 157, 158, 159
Job-enrichment specialists 36
Job security, loss of 72
Journalism, computers in 48
Journals, as sources of knowledge 26–27

Kantor (German mathematician) 55
Kell, Henry 65
Kennedy, President John F. 96
Kepler (German mathematician) 55
Kidney machine 127–28
Knowledge 135, 174
  acquisition of 11–13, 57
  classification of 55
  theory of 60
  updating of 26, 27

Labour Party 141
  rejects Lucas Plan 140
Labour Plan 178
Labour process 16, 19
Language 170–71
  scientific 10
Lathe 17
Latimer, Clive 118, 121
Leadership 162–63
  legitimacy of 134
Learning process 24

Learning through work 31
Leisure 30–31
Lenin 93, 164
Leninism 4
Leonardo da Vinci 59, 60
Libraries 50
Life-support system, portable 120
Light pen 19
Linguistic ability 70, 170–71
Livingstone, Ken 4
Llewelyn, Arthur 71
Lobell, Professor 51
London Airport 169
London Industrial Strategy 178
London Innovation Network 121, 144
Lordstown 106
Loughborough University 49
Low-energy housing 121
Lucas Aerospace 3–4, 36, 114–38, 154, 160, 163, 165, 168, 169, 177, 178
Lucas Electrical 122–24
Lucas Workers' Plan for Socially Useful Production 7, 139
Luddism 98
Lushai Hills effect 137, 176
Lyons 101

Machine-based systems 179
Machine intelligence 43
Macrae, Norman 26
Male values in science and technology 87
Management, philosophy of 133
'Man/Machine Systems Designing' 44
Manual work 9, 38, 73
Manufacturing industry, growth of 28
Mao Tse Tung 110
Margulies, Fred 43
Mariners' compasses 57
Marshall, Dennis 131
Martyn, Henry 78
Marx, Karl 52, 89, 90
Marxism 105
Masons' Guild 61
Massachusetts Institute of Technology 62
Material procurement 82
Mathematical modelling 51
Mathematics, development of 54–56
Mayer, Professor Tom 71
Medical checks 33
Medical profession,
  elitism of 145
  feudal mysticism of 120
Medicine, computers in 21
Mentally handicapped people performing repetitive tasks 130
Metalworkers' Union 132
Micrographics 76
Microplotter 17
Millwright, definition of 44
Milton Keynes Corporation 121–22
Mobility systems 145
Motivation 152
Multinational corporations 105, 115, 116, 129, 139, 157, 158, 159, 176
Mundy, Jack 164

National Health Service 103–104, 159, 162
National Liberation Front, Vietnam 110
Natural sciences 91
NEEB 77
Neocolonialism 125
Netherlands, unemployment in 96
Newspaper industry, computers in 48–49
New Technology Network 143
Newton, Isaac 52
New York 77, 114, 115, 120
New York City University 84
Noble, D. F. 172
NORA Report 32
North East London Polytechnic 118, 121, 126, 131
Northern Ireland 100, 105
North Sea oil pipelines, protection of 128
Norway 77
Nottingham University 102
Novices 14
Nuclear family 75

Index

Nuclear power 168–69
  strikebreaking capacity of 166–67
Nuremberg 55
Nyquist array 63, 136

Obsolescence 73, 74, 124
  planned 134
  rate of 25
Occupational growth areas 79
Olschki (German mathematician) 55
'On the Ordination of Pinnacles'
    60–61
Open University 118, 121
Operator commitment 152
Operator-interface software packages
    153
Operator role 150–51
Ostberg, O. 76
Overtime 73

Paper, reduction in volume of due to
    computerisation 75–76
Paris 58, 101
Parry Evans, Mike 120
Parsons 133
Part-programming languages 173
Pascal 71, 108
Pay structures 100–101
Peak-performance age 45–46
Pensioners, as discarded units of
    production 103
Performance 14, 15
Performance measurement 73
Philosophy of Manufactures 129
Planning 79
Plato 55
Political change 105
Politics, involvement of scientists in
    112
Pollution 105, 157–58, 162
Polytechnic productivity 82
Postural fatigue, due to use of VDUs
    76
Powell, John 84
Power-assisted bicycle, development
    of 145–46
Power generation 111
  all-purpose 124–26
Power relationships 140

Predictability 172
Print industry, computers in 48–49
Privacy, intrusion of 167
Production defects 35
Productivity 84, 172
Productivity deals 101
Proficiency 15
Profit 74, 105, 140, 157, 158, 177
Programming 79, 149, 150
Project management 69
Proletarianisation 73, 75, 94–97
Psychological factors of work 36
Puerto Ricans 98
Purdeys (Gunmakers) 64–65

Quality, concern for 64
Quality control 82
Quangos 98
Quantifiability 172
Quantification 91–92
Queen Mary College 118, 124

Radicalisation of scientific
    community 104
Rauner, Professor 63
Reagan, President Ronald 95
Real ale, production of 146
Real-time monitoring 153
Redundancy Payments Act 98
Regensburg 60, 61
Relativity 53
Religion 90
Repeatability 172
Research, illusions in 109
Resource planning and development
    82
Rest allowance 33–34
Rethel, Alfred Sohn 55
Retrospective logic 23–24
Revised reductionism 92
Rhythm of work 34
Rigal, Professor Jean-Louis 91
Road/rail vehicle 131, 146
  design of 126–27
Robotics 9, 36, 45, 102, 107, 128–29
  myth of 115
Robots as union members 42
Rolls-Royce 74, 133
Rome 34

Roriczer, Mathias 60–61
Rose, Steven 50
Rosenbrock, Professor Howard 63, 136, 137, 174–76, 178
Routines 73
Royal College of Art 131

St. Paul's Cathedral 66
Salzburg 61
Scarbrow, Ernie 127
Scheduling 149
Science,
　faith in 8
　hostility towards 116
　neutrality of 90–93, 133, 172, 176
Science Show 132
Scientific abuse 104
Scientific management 35–36, 79
Scientific methodology 91, 108
Seattle 95
Sellafield 168
Sex-discrimination legislation 88
Shakel, Professor 49
Shanghai 164
Shift work 73, 74, 75
Shop Management 78
Silver, Professor 92
Sinusoidal function 19
Skill 174
　acquisition of 13–16
　fragmentation of 73
Skill levels 67
Smith, Adam 78
Social awareness 51
Social control 93
Social Effects Committee of International Federation of Automatic Control 43
Social interaction 34
Social intercourse 50
Socialism 4, 93, 94, 162–63
Social life, disruption due to shift work 75
Socially useful production 154–56, 161
Social order 90
Social organisation, slow rate of change 132
Social responsibility 104, 105

Social sciences 92
Socrates 55
Soviet Union 93, 94, 162, 164, 166
Specialisation in education 45
Speer, Albert 176
Spina Bifida Association of Australia 119
Strasburg 61
Strathclyde University 71
Stress 39–40, 73
Stress factors of work 36
Strike, right to 167, 169
Strikes 33, 45, 80, 106, 110–111, 137
Structural analysis 19
Suffolk 121
Superannuation schemes 159
Superstition 90
Supplies procurement 82
Sweden 35, 76, 78, 106, 132, 137
Switzerland 57, 127
Symbiotic systems 63
Symbolic part-programming 174
Synge, J. M. 34

Tacit knowledge 11, 12, 13
Task-oriented time 34–35
Taylor, Frederick W. 2, 35, 38, 41, 66, 72, 73, 78, 80, 81, 93, 94, 101, 147, 161, 164, 172
Technical skills, acquisition of 64–65
Technological revolution 157
Technology,
　analysis of needed 178
　hostility towards 116
　neutrality of 172, 176
Technology Networks 154, 155
Telechiric devices 128–29, 147, 172
Telecommunication exchanges 29
Teletype 17
Thatcher, Margaret 98, 142, 166, 177
Therapeutic activities 31
Third World 8, 102, 119, 125, 155
Thorn EMI 133
Three Mile Island 166
Thring, Professor Meredith 118, 124
Time, task oriented 34–35
Times 48, 50
Trade Union Congress 131
Trade-union membership 73

Index

Trade-union movement 97, 117, 148–49, 161, 166, 167, 168–69, 177
  rejects Lucas Plan 140
Training 66–70
Training advisors 67–70
Transmission of data 22
Transport, development of new forms of 132
Transport and General Workers' Union 132, 140
Transport Network 143
Triumph Plant 32
Turning 173–74
Tyre, development of unpuncturable 146

Ulcers, due to shift work 75
UMIST 63, 148, 179
Unemployment 29, 30, 31, 32, 94–100, 115, 157, 159, 160, 161, 177
United States 28, 40, 41, 46, 48, 49, 76, 84, 91, 95, 96, 98, 106, 109, 135, 140, 145, 147, 158, 166
Universal power pack 125
University productivity 82
Ure, Andrew 129

Value judgements 39, 72, 73
VDUs and effect on health 75–77
Vickers 133
Victimisation for political views 167
Vienna 61
Vietnam War 95, 110
Visual discomfort, due to use of VDUs 76
Visual simulation techniques 23–24
Voluntary activity 161

Wainwright, H. 177
Wapping 149, 177
Warwick University 109
Washington 115
Washington University 83
Weinstock, Arnold 100, 116–17, 168
Weizenbaum, J. 62, 137
West Germany 32, 75, 95, 98, 123–24, 142, 147, 148, 150, 158, 167
  unemployment in 96
Whyte, William H. 109
Wiener, Norbert 50, 138
Wigner, Eugene 45
Willesden Hospital 104
Wilson, Harold 96, 117
Wind generation 124–26
Windloading, analysis of 19–21
Wittgenstein 154
Wolverhampton 119
Women, discrimination against 88–89
Work,
  importance of 30–31
  need for 36–37
  rhythm of 34
Work environment 73, 74
Workerless factory 153
Workers' control 165
Working week, length of 97
Work-measurement techniques 74
Work sharing 161
Workstudy 80
Workstudy 81
Work tempo 32–34, 73
Wottowski, Professor 63

Yalcs 80
Yoghurt production 146